Current Intelligence Bulletin 63

Occupational Exposure to Titanium Dioxide

DEPARTMENT OF HEALTH AND HUMAN SERVICES
Centers for Disease Control and Prevention
National Institute for Occupational Safety and Health

On the cover left to right: (1) Scanning electron microscopy (SEM) image of agglomerated particles of pigment-grade rutile TiO_2; (2) SEM image of agglomerated ultrafine-sized particles of rutile TiO_2. Images courtesy of Bill Fox, Altairnano, Inc., and Dr. Aleks Stefaniak and Dr. Mark Hoover, NIOSH Nanotechnology Field Research Team. Used with permission.

This document is in the public domain and may be freely copied or reprinted.

Disclaimer

Mention of any company or product does not constitute endorsement by the National Institute for Occupational Safety and Health (NIOSH). In addition, citations to Web sites external to NIOSH do not constitute NIOSH endorsement of the sponsoring organizations or their programs or products. Furthermore, NIOSH is not responsible for the content of these Web sites.

Ordering Information

To receive NIOSH documents or other information about occupational safety and health topics, contact NIOSH at

>   Telephone: 1–800–CDC–INFO (1–800–232–4636)
>   TTY: 1–888–232–6348
>   E-mail: cdcinfo@cdc.gov

or visit the NIOSH Web site at www.cdc.gov/niosh.

For a monthly update on news at NIOSH, subscribe to *NIOSH eNews* by visiting www.cdc.gov/niosh/eNews.

DHHS (NIOSH) Publication No. 2011–160

April 2011

SAFER • HEALTHIER • PEOPLE™

Foreword

The purpose of the Occupational Safety and Health Act of 1970 (Public Law 91–596) is to assure safe and healthful working conditions for every working person and to preserve our human resources. In this Act, the National Institute for Occupational Safety and Health (NIOSH) is charged with recommending occupational safety and health standards and describing exposures that are safe for various periods of employment, including (but not limited to) the exposures at which no worker will suffer diminished health, functional capacity, or life expectancy as a result of his or her work experience.

Current Intelligence Bulletins (CIBs) are issued by NIOSH to disseminate new scientific information about occupational hazards. A CIB may draw attention to a formerly unrecognized hazard, report new data on a known hazard, or disseminate information about hazard control. CIBs are distributed to representatives of academia, industry, organized labor, public health agencies, and public interest groups as well as to federal agencies responsible for ensuring the safety and health of workers.

Titanium dioxide (TiO_2), an insoluble white powder, is used extensively in many commercial products, including paint, cosmetics, plastics, paper, and food, as an anticaking or whitening agent. It is produced and used in the workplace in varying particle-size fractions, including fine and ultrafine sizes. The number of U.S. workers currently exposed to TiO_2 dust is unknown.

This NIOSH CIB, based on our assessment of the current available scientific information about this widely used material, (1) reviews the animal and human data relevant to assessing the carcinogenicity and other adverse health effects of TiO_2, (2) provides a quantitative risk assessment using dose-response information from the rat and human lung dosimetry modeling and recommended occupational exposure limits for fine and ultrafine (including engineered nanoscale) TiO_2, and (3) describes exposure monitoring techniques, exposure control strategies, and research needs. This report only addresses occupational exposures by inhalation, and conclusions derived here should not be inferred to pertain to nonoccupational exposures.

NIOSH recommends exposure limits of 2.4 mg/m³ for fine TiO_2 and 0.3 mg/m³ for ultrafine (including engineered nanoscale) TiO_2, as time-weighted average (TWA) concentrations for up to 10 hours per day during a 40-hour work week. NIOSH has determined that ultrafine TiO_2 is a potential occupational carcinogen but that there are insufficient data at this time to classify fine TiO_2 as a potential occupational carcinogen. However, as a precautionary step, NIOSH used all of the animal tumor response data when conducting dose-response modeling and determining separate

RELs for ultrafine and fine TiO_2. These recommendations represent levels that over a working lifetime are estimated to reduce risks of lung cancer to below 1 in 1,000. NIOSH realizes that knowledge about the health effects of nanomaterials is an evolving area of science. Therefore, NIOSH intends to continue dialogue with the scientific community and will consider any comments about nano-size titanium dioxide for future updates of this document. (Send comments to nioshdocket@cdc.gov.)

NIOSH urges employers to disseminate this information to workers and customers and requests that professional and trade associations and labor organizations inform their members about the hazards of occupational exposure to respirable TiO_2.

John Howard, M.D.
Director, National Institute for Occupational
   Safety and Health
Centers for Disease Control and Prevention

Executive Summary

In this Current Intelligence Bulletin, the National Institute for Occupational Safety and Health (NIOSH) reviews the animal and human data relevant to assessing the carcinogenicity of titanium dioxide (TiO_2) (Chapters 2 and 3), presents a quantitative risk assessment using dose-response data in rats for both cancer (lung tumors) and noncancer (pulmonary inflammation) responses and extrapolation to humans with lung dosimetry modeling (Chapter 4), provides recommended exposure limits (RELs) for fine and ultrafine (including engineered nanoscale) TiO_2 (Chapter 5), describes exposure monitoring techniques and exposure control strategies (Chapter 6), and discusses avenues of future research (Chapter 7). This report only addresses occupational exposures by inhalation, and conclusions derived here should not be inferred to pertain to nonoccupational exposures.

TiO_2 (Chemical Abstract Service [CAS] Number 13463–67–7) is a noncombustible, white, crystalline, solid, odorless powder. TiO_2 is used extensively in many commercial products, including paints and varnishes, cosmetics, plastics, paper, and food as an anticaking or whitening agent. Production in the United States was an estimated 1.45 million metric tons per year in 2007 [DOI 2008]. The number of U.S. workers currently exposed to TiO_2 dust is not available.

TiO_2 is produced and used in the workplace in varying particle size fractions including fine (which is defined in this document as all particle sizes collected by respirable particle sampling) and ultrafine (defined as the fraction of respirable particles with a primary particle diameter of <0.1 μm [<100 nm]). Particles <100 nm are also defined as nanoparticles.

The Occupational Safety and Health Administration (OSHA) permissible exposure limit for TiO_2 is 15 mg/m³, based on the airborne mass fraction of total TiO_2 dust (Chapter 1). In 1988, NIOSH recommended that TiO_2 be classified as a potential occupational carcinogen and that exposures be controlled as low as feasible [NIOSH 2002]. This recommendation was based on the observation of lung tumors (nonmalignant) in a chronic inhalation study in rats at 250 mg/m³ of fine TiO_2 [Lee et al. 1985, 1986a] (Chapter 3).

Later, a 2-year inhalation study showed a statistically significant increase in lung cancer in rats exposed to ultrafine TiO_2 at an average concentration of 10 mg/m³ [Heinrich et al. 1995]. Two recent epidemiologic studies have not found a relationship between exposure to total or respirable TiO_2 and lung cancer [Fryzek et al. 2003; Boffetta et al. 2004], although an elevation in lung cancer mortality was observed among male TiO_2 workers in the latter study when compared to the general population (standardized mortality ratio [SMR] 1.23; 95% confidence interval [CI] = 1.10–1.38) (Chapter 2). However, there was no indication of an exposure-response relationship in that study. Nonmalignant respiratory disease mortality was not increased significantly ($P <0.05$) in any of the epidemiologic studies.

In 2006, the International Agency for Research on Cancer (IARC) reviewed TiO_2 and concluded that there was sufficient evidence of carcinogenicity in experimental animals and inadequate evidence of carcinogenicity in humans (Group 2B), "possibly carcinogenic to humans" [IARC 2010].

TiO_2 and other poorly soluble, low-toxicity (PSLT) particles of fine and ultrafine sizes show a consistent dose-response relationship for adverse pulmonary responses in rats, including persistent pulmonary inflammation and lung tumors, when dose is expressed as particle surface area. The higher mass-based potency of ultrafine TiO_2 compared to fine TiO_2 is associated with the greater surface area of ultrafine particles for a given mass. The NIOSH RELs for fine and ultrafine TiO_2 reflect this mass-based difference in potency (Chapter 5). NIOSH has reviewed and considered all of the relevant data related to respiratory effects of TiO_2. This includes results from animal inhalation studies and epidemiologic studies. NIOSH has concluded that TiO_2 is not a direct-acting carcinogen, but acts through a secondary genotoxicity mechanism that is not specific to TiO_2 but primarily related to particle size and surface area. The most relevant data for assessing the health risk to workers are results from a chronic animal inhalation study with ultrafine (<100 nm) TiO_2 in which a statistically significant increase in adenocarcinomas was observed [Heinrich et al. 1995]. This is supported by a pattern of TiO_2 induced responses that include persistent pulmonary inflammation in rats and mice [Everitt et al. 2000; Bermudez et al. 2004] and cancer responses for PSLT particles related to surface area. Therefore, on the basis of the study by Heinrich et al. [1995] and the pattern of pulmonary inflammatory responses, NIOSH has determined that exposure to ultrafine TiO_2 should be considered a potential occupational carcinogen.

For fine size (pigment grade) TiO_2 (>100 nm), the data on which to assess carcinogenicity are limited. Generally, the epidemiologic studies for fine TiO_2 are inconclusive because of inadequate statistical power to determine whether they replicate or refute the animal dose-response data. This is consistent for carcinogens of low potency. The only chronic animal inhalation study [Lee et al. 1985], which demonstrated the development of lung tumors (bronchioalveolar adenomas) in response to inhalation exposure of rats to fine sized TiO_2 did so at a dose of 250 mg/m^3 but not at 10 or 50 mg/m^3. The absence of lung tumor development for fine TiO_2 was also reported by Muhle et al. [1991] in rats exposed at 5 mg/m^3. However, the responses observed in animal studies exposed to ultrafine and fine TiO_2 are consistent with a continuum of biological response to TiO_2 that is based on particle surface area. In other words, all the rat tumor response data on inhalation of TiO_2 (ultrafine and fine) fit on the same dose-response curve when dose is expressed as total particle surface area in the lungs. However, exposure concentrations greater than 100 mg/m^3 are generally not considered acceptable inhalation toxicology practice today. Consequently, in a weight-of-evidence analysis, NIOSH questions the relevance of the 250 mg/m^3 dose for classifying exposure to TiO$_2$ as a carcinogenic hazard to workers and therefore, concludes that there are insufficient data at this time to classify fine TiO$_2$ as a potential occupational carcinogen. Although data are insufficient on the cancer hazard for fine TiO$_2$, the tumor-response data are consistent with that observed for ultrafine TiO$_2$ when converted to a particle surface area metric. Thus to be cautious, NIOSH used all of the animal tumor response data when conducting dose-response modeling and determining separate RELs for ultrafine and fine TiO$_2$.

NIOSH also considered the crystal structure as a modifying factor in TiO$_2$ carcinogenicity and inflammation. The evidence for crystal-dependent toxicity is from observed differences in reactive oxygen species (ROS) generated on the surface of TiO$_2$ of different crystal structures (e.g., anatase, rutile, or mixtures) in cell-free systems, with differences in cytotoxicity in *in vitro* studies [Kawahara et al. 2003; Kakinoki et al. 2004; Behnajady et al. 2008; Jiang et al. 2008, Sayes et al. 2006] and with greater inflammation and cell proliferation at early time points following intratracheal instillation in rats [Warheit et al. 2007]. However, when rats were exposed to TiO$_2$ in subchronic inhalation studies, no difference in pulmonary inflammation response to fine and ultrafine TiO$_2$ particles of different crystal structure (i.e., 99% rutile vs. 80% anatase/20% rutile) was observed once dose was adjusted for particle surface area [Bermudez et al. 2002, 2004]; nor was there a difference in the lung tumor response in the chronic inhalation studies in rats at a given surface area dose of these fine and ultrafine particles (i.e., 99% rutile vs. 80% anatase/20% rutile) [Lee et al. 1985; Heinrich et al. 1995]. Therefore, NIOSH concludes that the scientific evidence supports surface area as the critical metric for occupational inhalation exposure to TiO$_2$.

NIOSH also evaluated the potential for coatings to modify the toxicity of TiO$_2$, as many industrial processes apply coatings to TiO$_2$ particles. TiO$_2$ toxicity has been shown to increase after coating with various substances [Warheit et al. 2005]. However, the toxicity of TiO$_2$ has not been shown to be attenuated by application of coatings. NIOSH concluded that the TiO$_2$ risk assessment could be used as a reasonable floor for potential toxicity, with the notion that toxicity may be substantially increased by particle treatment and process modification. These findings are based on the studies in the scientific literature and may not apply to other formulations, surface coatings, or treatments of TiO$_2$ for which data were not available. An extensive review of the risks of coated TiO$_2$ particles is beyond the scope of this document.

NIOSH recommends airborne exposure limits of 2.4 mg/m^3 for fine TiO$_2$ and 0.3 mg/m^3 for ultrafine (including engineered nanoscale) TiO$_2$, as time-weighted average (TWA) concentrations for up to 10 hr/day during a 40-hour work week. These recommendations represent levels that over a working lifetime are estimated to reduce risks of lung cancer to below 1 in 1,000. The recommendations are based on using chronic inhalation studies in rats to predict lung tumor risks in humans.

In the hazard classification (Chapter 5), NIOSH concludes that the adverse effects of inhaling TiO_2 may not be material-specific but appear to be due to a generic effect of PSLT particles in the lungs at sufficiently high exposure. While NIOSH concludes that there is insufficient evidence to classify fine TiO_2 as a potential occupational carcinogen, NIOSH is concerned about the potential carcinogenicity of ultrafine and engineered nanoscale TiO_2 if workers are exposed at the current mass-based exposure limits for respirable or total mass fractions of TiO_2. NIOSH recommends controlling exposures as low as possible, below the RELs. Sampling recommendations based on current methodology are provided (Chapter 6).

Although sufficient data are available to assess the risks of occupational exposure to TiO_2, additional research questions have arisen. There is a need for exposure assessment for workplace exposure to ultrafine TiO_2 in facilities producing or using TiO_2. Other research needs include evaluation of the (1) exposure-response relationship of TiO_2 and other PSLT particles and human health effects, (2) fate of ultrafine particles in the lungs and the associated pulmonary responses, and (3) effectiveness of engineering controls for controlling exposures to fine and ultrafine TiO_2. (Research needs are discussed further in Chapter 7).

Contents

Foreword	iii
Executive Summary	v
Abbreviations	xii
Acknowledgments	xv
1 Introduction	1
1.1 Composition	1
1.2 Uses	2
1.3 Production and number of workers potentially exposed	2
1.4 Current exposure limits and particle size definitions	3
2 Human Studies	7
2.1 Case Reports	7
2.2 Epidemiologic Studies	8
2.2.1 Chen and Fayerweather [1988]	8
2.2.2 Fryzek et al. [2003]	17
2.2.3 Boffetta et al. [2001]	18
2.2.4 Boffetta et al. [2004]	19
2.2.5 Ramanakumar et al. [2008]	20
2.3 Summary of Epidemiologic Studies	21
3 Experimental Studies in Animals and Comparison to Humans	23
3.1 In Vitro Studies	23
3.1.1 Genotoxicity and Mutagenicity	23
3.1.2 Oxidant Generation and Cytotoxicity	24
3.1.3 Effects on Phagocytosis	24
3.2 In Vivo Studies in Rodent Lungs	25
3.2.1 Intratracheal Instillation	25
3.2.2 Acute or Subacute Inhalation	29
3.2.3 Short-term Inhalation	30
3.2.4 Subchronic Inhalation	31
3.2.5 Chronic Inhalation	33
3.3 In Vivo Studies: Other Routes of Exposure	35
3.3.1 Acute Oral Administration	35

  3.3.2 Chronic Oral Administration 35
  3.3.3 Intraperitoneal Injection 36
 3.4 Particle-Associated Lung Disease Mechanisms 36
  3.4.1 Role of Pulmonary Inflammation 36
  3.4.2 Dose Metric and Surface Properties 37
 3.5 Particle-Associated Lung Responses 42
  3.5.1 Rodent Lung Responses to Fine and Ultrafine TiO_2 42
  3.5.2 Comparison of Rodent and Human Lung Responses to
    PSLT including TiO_2. 43
 3.6 Rat Model in Risk Assessment of Inhaled Particles 48

4 Quantitative Risk Assessment 51
 4.1 Data and Approach 51
 4.2 Methods. 51
  4.2.1 Particle Characteristics 51
  4.2.2 Critical Dose 53
  4.2.3 Estimating Human Equivalent Exposure 53
  4.2.4 Particle Dosimetry Modeling 54
 4.3 Dose-Response Modeling of Rat Data and Extrapolation
   to Humans. 54
  4.3.1 Pulmonary Inflammation 54
  4.3.2 Lung Tumors. 59
  4.3.3 Alternate Models and Assumptions. 63
  4.3.4 Mechanistic Considerations 67
 4.4 Quantitative Comparison of Risk Estimates from Human and
   Animal Data 68
 4.5 Possible Bases for an REL 68
  4.5.1 Pulmonary Inflammation 68
  4.5.2 Lung Tumors. 69
  4.5.3 Comparison of Possible Bases for an REL. 70

5 Hazard Classification and Recommended Exposure Limits. 73
 5.1 Hazard Classification 73
  5.1.1 Mechanistic Considerations 74
  5.1.2 Limitations of the Rat Tumor Data 75
  5.1.3 Cancer Classification in Humans. 76
 5.2 Recommended Exposure Limits 77

6 Measurement and Control of TiO_2 Aerosol in the Workplace 79
 6.1 Exposure Metric. 79
 6.2 Exposure Assessment. 80
 6.3 Control of Workplace Exposures to TiO_2 80

7 Research Needs 85
    7.1 Workplace Exposures and Human Health 85
    7.2 Experimental Studies. 85
    7.3 Measurement, Controls, and Respirators 85

References 87

Appendices
    A. Statistical Tests of the Rat Lung Tumor Models. 105
    B. Threshold Model for Pulmonary Inflammation in Rats 111
    C. Comparison of Rat- and Human-based Excess Risk Estimates
        for Lung Cancer Following Chronic Inhalation of TiO_2. 113
    D. Calculation of Upper Bound on Excess Risk of Lung Cancer
        in an Epidemiologic Study of Workers Exposed to TiO_2 117

Abbreviations

ACGIH	American Conference of Governmental Industrial Hygienists
BAL	bronchoalveolar lavage
BALF	bronchoalveolar lavage fluid
BAP	benzo(a)pyrene
$BaSO_4$	barium sulfate
BET	Brunauer, Emmett, and Teller
BMD	benchmark dose
BMDL	benchmark dose lower bound
BMDS	benchmark dose software
°C	degree(s) Celsius
CAS	Chemical Abstract Service
CFR	Code of Federal Regulations
CI	confidence interval
cm	centimeter(s)
DNA	deoxyribonucleic acid
E	expected
EDS	energy dispersive spectroscopy
g	gram(s)
g/cm^3	grams per cubic centimeter
g/ml	gram per milliliter
GSD	geometric standard deviation
hprt	hypoxanthine-guanine phosphoribosyl transferase
hr	hour(s)
IARC	International Agency for Research on Cancer
ICRP	International Commission on Radiological Protection
IR	incidence ratio
IT	intratracheal instillation
kg	kilogram
L	liter(s)
LCL	lower confidence limit
LDH	lactate dehydrogenase
m	meter(s)
MA	model average

MAK	Federal Republic of Germany maximum concentration value in the workplace
MCEF	mixed cellulose ester filter
mg	milligram(s)
mg/kg	milligram per kilogram body weight
mg/m^3	milligrams per cubic meter
mg/m^3 • yr	milligrams per cubic meter times years
min	minute(s)
ml	milliliter(s)
ML	maximum likelihood
MLE	maximum likelihood estimate
mm	millimeter(s)
MMAD	mass median aerodynamic diameter
MPPD	multiple-path particle dosimetry
n	number
NAICS	North American Industry Classification System
NCI	National Cancer Institute
NIOSH	National Institute for Occupational Safety and Health
nm	nanometer(s)
NOAEL	no-observed-adverse-effect level
O	observed
OR	odds ratio
OSHA	Occupational Safety and Health Administration
P	probability
PBS	phosphate buffered saline
PEL	permissible exposure limit
PH	proportional hazards
PKT	pigmentary potassium titinate
PMN	polymorphonuclear leukocytes
PNOR/S	particles not otherwise regulated or specified
PNOR	particles not otherwise regulated
PNOS	particles not otherwise specified
PSLT	poorly soluble, low toxicity
REL	recommended exposure limit
ROS	reactive oxygen species
RNS	reactive nitrogen species
RR	relative risk
SiO$_2$	silicon dioxide
SMR	standardized mortality ratio
TEM	transmission electron microscopy

$TiCl_4$	titanium tetrachloride
TiO_2	titanium dioxide
TWA	time-weighted average
UCL	upper confidence limit
U.K.	United Kingdom
UV	ultraviolet
U.S.	United States
wk	week(s)
μg	microgram(s)
μm	micrometer(s)
%	percent

Acknowledgments

This Current Intelligence Bulletin (CIB) was prepared by the Education and Information Division (EID), Paul Schulte, Director; Risk Evaluation Branch, Christine Sofge, Chief; Document Development Branch, T.J. Lentz, Chief. Faye Rice (EID) managed the preparation of the final CIB and the responses to external review comments. The document was authored by the EID Titanium Dioxide Document Development Team and interdivisional authors who developed first drafts of some chapters and sections.

EID Document Development Team

David Dankovic and Eileen Kuempel (primary authors), and in alphabetical order, Charles Geraci, Stephen Gilbert, Faye Rice, Paul Schulte, Randall Smith, Christine Sofge, Matthew Wheeler, Ralph Zumwalde

Division of Applied Research and Technology (DART)

Andrew Maynard (currently with the University of Michigan School of Public Health, Risk Science Center)

Division of Respiratory Disease Studies (DRDS)

Michael Attfield, Germania Pinheiro (currently with the National Center for HIV/AIDS, Viral Hepatitis, STD, and TB Prevention)

Division of Surveillance, Hazard Evaluations, and Field Studies (DSHEFS)

Avima Ruder

Health Effects Laboratory Division (HELD)

Ann Hubbs

Cross-divisional team to evaluate data on carcinogenicity of TiO_2

David Dankovic (EID)
Heinz Ahlers (EID), (currently with the National Personal Protective Technology Laboratory [NPPTL])
Vincent Castranova (HELD)
Eileen Kuempel (EID)
Dennis Lynch (DART)
Avima Ruder (DSHEFS)
Mark Toraason (DART)
Val Vallyathan (HELD)
Ralph Zumwalde (EID)

Other NIOSH reviewers who provided critical feedback important to the preparation of the document

Paul Middendorf (Office of the Director (OD), Sid Soderholm (OD), Jimmy Stephens (OD), Frank Hearl (OD), John Piacentino (OD), Roger Rosa (OD), John Decker (OD), Paul Baron (DART), Cynthia Striley (DART), G. Scott Earnest (DART), Jennifer Topmiller (DART), Patricia Sullivan (DRDS), John McKernan (DSHEFS), Mary Schubauer-Berigan (DSHEFS), Richard Niemeier (EID), Roland BerryAnn (NPPTL)

EID Editorial and Document Assistance

Devin Baker, Anne Hamilton, Norma Helton, Laurel Jones, John Lechliter, Alma McLemore, Jessica Porco, Brenda Proffitt, Cathy Rotunda, Stella Stephens, Jane Weber, Linda Worley

Document Design and Layout

Vanessa B. Williams (EID)

NIOSH acknowledges the contribution of the following scientists who provided data and/or additional information on their studies to aid the NIOSH risk assessment and comparative analysis: Dr. Tran of IOM, Dr. Bermudez of the Hamner Institute, Dr. Fryzek, currently at MedImmune, and Dr. Boffetta, currently at the Tisch Cancer Institute at Mount Sinai School of Medicine.

NIOSH expresses appreciation to the following independent, external reviewers for providing comments that contributed to the development of this document:

External Expert Peer Review Panel

Chao W. Chen, Ph.D.
Senior Statistician
National Center for Environmental Assessment
U.S. Environmental Protection Agency

Harvey Clewell, Ph.D.
Director, Center for Human Health Assessment
Centers for Health Research
Chemical Industry Institute of Toxicology (CIIT)

Prof. Dr. med. Helmut Greim
Institute of Toxicology and Environmental Hygiene
Technical University of Munich

Franklin E. Mirer, Ph.D., CIH
Director, Health and Safety Department
International Union, UAW (at time of review)
Current title and affiliation:
Professor
Environmental and Occupational Health Sciences
Urban Public Health Program
Hunter School of Health Sciences

Jonathan M. Samet, MD, MS
Chairman, Department of Epidemiology
Bloomberg School of Public Health
Johns Hopkins University

1 Introduction

1.1 Composition

Titanium dioxide (TiO$_2$), Chemical Abstract Service [CAS] (CAS Number 13463–67–7), is a noncombustible, white, crystalline, solid, odorless powder [NIOSH 2002; ACGIH 2001a]. TiO$_2$ is insoluble in water, hydrochloric acid, nitric acid, or alcohol, and it is soluble in hot concentrated sulfuric acid, hydrogen fluoride, or alkali [ACGIH 2001a]. TiO$_2$ has several naturally occurring mineral forms, or polymorphs, which have the same chemical formula and different crystalline structure. Common TiO$_2$ polymorphs include rutile (CAS Number 1317–80–2) and anatase (CAS Number 1317–70–0). While both rutile and anatase belong to the tetragonal crystal system, rutile has a denser arrangement of atoms (Figure 1).

At temperatures greater than 915°C, anatase reverts to the rutile structure [http://mineral.galleries.com/minerals/oxides/anatase/anatase.htm]. The luster and hardness of anatase and rutile are also similar, but the cleavage differs. The density (specific gravity) of rutile is 4.25 g/ml [http://webmineral.com/data/Rutile.shtml], and that of anatase is 3.9 g/ml [http://webmineral.com/data/Anatase.shtml]. Common impurities in rutile include iron, tantalum, niobium, chromium, vanadium, and tin [http://www.mindat.org/min-3486.html], while those in anatase include iron, tin, vanadium, and niobium [http://www.mindat.org/min-213.html].

The sulfate process and the chloride process are two main industrial processes that produce TiO$_2$

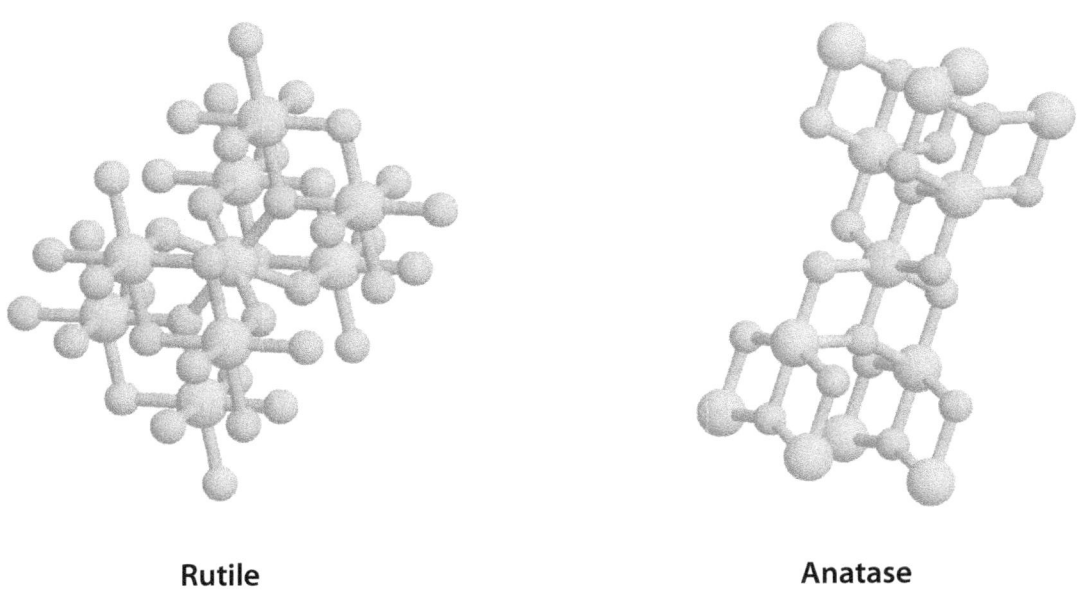

Rutile **Anatase**

Figure 1. Rutile and anatase TiO$_2$ crystal structure (Image courtesy of Cynthia Striley, NIOSH)

pigment [IARC 1989; Boffetta et al. 2004]. In the sulfate process, anatase or rutile TiO_2 is produced by digesting ilmenite (iron titanate) or titanium slag with sulfuric acid. In the chloride process, natural or synthetic rutile is chlorinated at temperatures of 850 to 1,000°C [IARC 1989], and the titanium tetrachloride (TiCl4) is converted to the rutile form by vapor-phase oxidation [Lewis 1993]. Both anatase and rutile are used as white pigment. Rutile TiO_2 is the most commonly used white pigment because of its high refractive index and relatively low absorption of light [Wicks 1993]. Anatase is used for specialized applications (e.g., in paper and fibers). TiO_2 does not absorb visible light, but it strongly absorbs ultraviolet (UV) radiation. Commercial rutile TiO_2 is prepared with an average particle size of 0.22 µm to 0.25 µm [Wicks 1993]. Pigment-grade TiO_2 refers to anatase and rutile pigments with a median particle size that usually ranges from 0.2 µm to 0.3 µm [Aitken et al. 2004]. Particle size is an important determinant of the properties of pigments and other final products [Wicks 1993].

1.2 Uses

TiO_2 is used mainly in paints, varnishes, lacquer, paper, plastic, ceramics, rubber, and printing ink. TiO_2 is also used in welding rod coatings, floor coverings, catalysts, coated fabrics and textiles, cosmetics, food colorants, glassware, pharmaceuticals, roofing granules, rubber tire manufacturing, and in the production of electronic components and dental impressions [Lewis 1993; ACGIH 2001a; IARC 1989; DOI 2005]. Both the anatase and rutile forms of TiO_2 are semiconductors [Egerton 1997]. TiO_2 white pigment is widely used due to its high refractive index. Since the 1960s, TiO_2 has been coated with other materials (e.g., silica, alumina) for commercial applications [Lee et al. 1985].

1.3 Production and Number of Workers Potentially Exposed

An estimate of the number of U.S. workers currently exposed to TiO_2 dust is not available. The only current information is an unreferenced estimate submitted by industry to NIOSH in response to request for public comment on the draft document. Industry estimates that the number of U.S. workers in the "so-called 'white' end of TiO_2 production plants" is "approximately 1,100 workers nationwide" and that there is no reliable estimate of the number of workers involved in the "initial compounding of downstream products" [American Chemistry Council 2006].

In 2007, an estimated 1.45 million metric tons of TiO_2 pigment were produced by four U.S. companies at eight facilities in seven states that employed an estimated 4,300 workers (jobs not described) [DOI 2008]. The paint (includes varnishes and lacquers), plastic and rubber, and paper industries accounted for an estimated 95% of TiO_2 pigment used in the United States in 2004 [DOI 2005]. In 2006, the U.S. Bureau of Labor Statistics estimated that there were about 68,000 U.S. workers in all occupations (excluding self-employed workers) in paint, coating, and adhesive manufacturing (North American Industry Classification System [NAICS] code 325500), 803,000 in plastics and rubber products manufacturing (NAICS code 326000), and about 138,000 employed in pulp, paper, and paperboard mills (NAICS code 322100) [BLS 2006]. In 1991, TiO_2 was the 43rd highest-volume chemical produced in the United States [Lewis 1993].

1.4 Current Exposure Limits and Particle Size Definitions

Occupational exposure to TiO$_2$ is regulated by the Occupational Safety and Health Administration (OSHA) under the permissible exposure limit (PEL) of 15 mg/m^3 for TiO$_2$ as total dust (8-hr time-weighted average [TWA] concentration) [29 CFR* 1910.1000; Table Z–1]. The OSHA PEL for particles not otherwise regulated (PNOR) is 5 mg/m^3 as respirable dust [29 CFR 1910.1000; Table Z–1]. These and other exposure limits for TiO$_2$ and PNOR or particles not otherwise specified (PNOS) are listed in Table 1. PNOR/S are defined as all inert or nuisance dusts, whether mineral, inorganic or organic, not regulated specifically by substance name by OSHA (PNOR) or classified by the American Conference of Governmental Industrial Hygienists (ACGIH)(PNOS). The same exposure limits are often given for TiO$_2$ and PNOR/PNOS (Table 1). OSHA definitions for the total and respirable particle size fractions refer to specific sampling methods and devices [OSHA 2002], while the maximum concentration value in the workplace (MAK) and the ACGIH definitions for respirable and inhalable particle sizes are based on the internationally developed definitions of particle size selection sampling [CEN 1993; ISO 1995; ACGIH 1984, 1994]. NIOSH also recommends the use of the international definitions [NIOSH 1995].

Aerodynamic diameter affects how a particle behaves in air and determines the probability of deposition at locations within the respiratory tract. Aerodynamic diameter is defined as the diameter of a spherical particle that has the same settling velocity as a particle with a density of 1 g/cm^3 (the density of a water droplet) [Hinds 1999].

"Respirable" is defined as the aerodynamic size of particles that, when inhaled, are capable of depositing in the gas-exchange (alveolar) region of the lungs [ICRP 1994]. Sampling methods have been developed to estimate the airborne mass concentration of respirable particles [CEN 1993; ISO 1995; ACGIH 1994; NIOSH 1998].

"Fine" is defined in this document as all particle sizes that are collected by respirable particle sampling (i.e., 50% collection efficiency for particles of 4 µm, with some collection of particles up to 10 µm). Fine is sometimes used to refer to the particle fraction between 0.1 µm and approximately 3 µm [Aitken et al. 2004], and to pigment-grade TiO$_2$ [e.g., Lee et al. 1985]. The term "fine" has been replaced by "respirable" by some organizations, e.g., MAK [DFG 2000], which is consistent with international sampling conventions [CEN 1993; ISO 1995].

"Ultrafine" is defined as the fraction of respirable particles with primary particle diameter <0.1 µm (<100 nm), which is a widely used definition. Particles <100 nm are also defined as nanoparticles. A primary particle is defined as the smallest identifiable subdivision of a particulate system [BSI 2005]. Additional methods are needed to determine if an airborne respirable particle sample includes ultrafine TiO$_2$ (Chapter 6). In this document, the terms fine and respirable are used interchangeably to retain both the common terminology and the international sampling convention.

In 1988, NIOSH classified TiO$_2$ as a potential occupational carcinogen and did not establish a recommended exposure limit (REL) for TiO$_2$ [NIOSH 2002]. This classification was based on the observation that TiO$_2$ caused lung tumors

---

*See CFR in references.

Table 1. Occupational exposure limits and guidelines for TiO_2* and PNOR/S

Agency	TiO$_2$ Single-shift TWA (mg/m³)	TiO$_2$ Comments	PNOR/S Single-shift TWA (mg/m³)	PNOR/S Comments
NIOSH [2002]†	—	Potential human carcinogen	—	—
OSHA	15	Total‡	15 5	Total Respirable
ACGIH [2001a, 2001b, 2005, 2009]	10	Category A4 (not classifiable as a human carcinogen); TiO$_2$ is under study by ACGIH [ACGIH 2009].	10§ 3§	Inhalable Respirable
MAK†† [DFG 2000, 2008]	—	Inhalable fraction except for ultrafine particles; suspected carcinogen (MAK Category 3A)	4 1.5	Inhalable Respirable

*Abbreviations: ACGIH = American Conference of Governmental Industrial Hygienists; MAK = Federal Republic of Germany Maximum Concentration Values in the Workplace; NIOSH = National Institute for Occupational Safety and Health; OSHA = Occupational Safety and Health Administration; PNOR/S = particles not otherwise regulated or specified; TiO$_2$ = titanium dioxide; TWA = time-weighted average; TLV® = threshold limit value
†Recommendations in effect before publication of this document.
‡*Total*, *inhalable*, and *respirable* refer to the particulate size fraction, as defined by the respective agencies.
§PNOS guideline (too little evidence to assign TLV®). Applies to particles without applicable TLV®, insoluble or poorly soluble, low toxicity (PSLT) [ACGIH 2005, 2009]. Inorganic only; and for particulate matter containing no asbestos and < 1% crystalline silica [ACGIH 2001b].
††MAK values are long-term averages. Single shift excursions are permitted within a factor of 2 of the MAK value. The TiO$_2$ MAK value has been withdrawn and pregnancy risk group C is not applicable [DFG 2008].

in rats in a long-term, high-dose bioassay [Lee et al. 1985]. NIOSH concluded that the results from this study met the criteria set forth in the OSHA cancer policy (29 CFR Part 1990, Identification, Classification, and Regulation of Carcinogens) by producing tumors in a long-term mammalian bioassay. The International Agency for Research on Cancer (IARC) formerly classified TiO_2 in Group 3, with limited evidence of animal carcinogenicity and inadequate evidence for human carcinogenicity [IARC 1989]. In 2006, IARC classified TiO_2 in Group 2B, with sufficient evidence of carcinogenicity in experimental animals and inadequate evidence for human carcinogenicity and an overall evaluation of "possibly carcinogenic to humans (Group 2B)" [IARC 2010]. The scientific evidence pertaining to hazard classification and exposure limits for TiO_2 is reviewed and evaluated in this document.

2 Human Studies

2.1 Case Reports

Case reports can provide information about the potential health effects of exposure to titanium dioxide (TiO_2) that may lead to formal epidemiologic studies of a relationship between occupational exposure and observed cases.

A few case reports described adverse health effects in workers with potential TiO_2 exposure. These effects included adenocarcinoma of the lung and TiO_2-associated pneumoconiosis in a male TiO_2 packer with 13 years of potential dust exposure and a 40-year history of smoking [Yamadori et al. 1986]. Pulmonary fibrosis or fibrotic changes and alveolar macrophage responses were identified by thoracotomy or autopsy tissue sampling in three workers with 6 to 9 years of dusty work in a TiO_2 factory. No workplace exposure data were reported. Two workers were "moderate" or "heavy" smokers (pack-years not reported) and smoking habits were not reported for the other worker [Elo et al. 1972]. Small amounts of silica were present in all three lung samples, and significant nickel was present in the lung tissue of the autopsied case. Exposure was confirmed using sputum samples that contained macrophages with high concentrations of titanium 2 to 3 years after their last exposure [Määttä and Arstila 1975]. Titanium particles were identified in the lymph nodes of the autopsied case. The lung concentrations of titanium were higher than the lung concentration range of control autopsy specimens from patients not exposed to TiO_2 (statistical testing and number of controls not reported).

Moran et al. [1991] presented cases of TiO_2 exposure in four males and two females. However, occupation was unknown for one male and one female, and the lung tissue of one worker (artist/painter) was not examined (skin biopsy of arm lesions was performed). Smoking habits were not reported. Diffuse fibrosing interstitial pneumonia, bronchopneumonia, and alveolar metaplasia were reported in three male patients (a TiO_2 worker, a painter, and a paper mill worker) with lung-deposited TiO_2 (rutile) and smaller amounts of tissue-deposited silica [Moran et al. 1991]. Titanium was also identified in the liver, spleen, and one peribronchial lymph node of the TiO_2 worker, and talc was identified in the lungs of that patient and the paper mill worker.

A case of pulmonary alveolar proteinosis (i.e., deposition of proteinaceous and lipid material within the airspaces of the lung) was reported in a worker employed for more than 25 years as a painter, with 8 years of spray painting experience. He smoked two packs of cigarettes per day until he was hospitalized. Titanium was the major type of metallic particle found in his lung tissues [Keller et al. 1995].

According to a four-sentence abstract from the Toxic Exposure Surveillance System (TESS) of the American Association of Poison Control Centers, death occurred suddenly in a 26-year-old male worker while pressure-cleaning inside a tank containing TiO_2; death was "felt to be due to inhalation of this particulate chemical" [Litovitz et al. 2002; Litovitz 2004]. There was no other information about the cause of death or indication that an autopsy was conducted.

TESS data are used for hazard identification, education, and training [Litovitz et al. 2002; Litovitz 2004].

In pathology studies of TiO_2 workers, tissue-deposited titanium was often used to confirm exposure. In many cases, titanium, rather than TiO_2, was identified in lung tissues; the presence of TiO_2 was inferred when a TiO_2-exposed worker had pulmonary deposition of titanium (e.g., Ophus et al. [1979]; Rode et al. [1981]; Määttä and Arstila [1975]; Elo et al. [1972]; Humble et al. [2003]). In other case reports, X-ray crystallography identified TiO_2 (i.e., anatase) in tissue digests [Moran et al. 1991], and X-ray diffraction distinguished rutile from anatase [Rode et al. 1981]. Similarly, with the exception of one individual in whom talc was identified [Moran et al. 1991], pathology studies (i.e., Elo et al. [1972]; Moran et al. [1991]) identified the silica as "SiO_2" (silicon dioxide) or "silica" in tissue and did not indicate whether it was crystalline or amorphous.

In summary, few TiO_2-related health effects were identified in case reports. None of the case reports provided quantitative industrial hygiene information about workers' TiO_2 dust exposure. Lung particle analyses indicated that workers exposed to respirable TiO_2 had particle retention in their lungs that included titanium, silica (form not specified), and other minerals sometimes years after cessation of exposure. The chronic tissue reaction to lung-deposited titanium is distinct from chronic silicosis. Most cases of tissue-deposited titanium presented with a local macrophage response with associated fibrosis that was generally mild, but of variable severity, at the site of deposition. More severe reactions were observed in a few cases. The prevalence of similar histopathologic responses in other TiO_2-exposed populations is not known. The effects of concurrent or sequential exposure to carcinogenic particles, such as crystalline silica, nickel, and tobacco smoke, were not determined.

2.2 Epidemiologic Studies

A few epidemiologic studies have evaluated the carcinogenicity of TiO_2 in humans; they are described here and in Table 2–1. Epidemiologic studies of workers exposed to related compounds, such as $TiCl_4$ or titanium metal dust (i.e., Fayerweather et al. [1992] and Garabrant et al. [1987]) were not included because those compounds may have properties and effects that differ from those of TiO_2 and discussion of those differences is beyond the scope of this document.

2.2.1 Chen and Fayerweather [1988]

Chen and Fayerweather [1988] conducted a mortality, morbidity, and nested case-control study of 2,477 male wage-grade workers employed for more than 1 year before January 1, 1984 in two TiO_2 production plants in the United States. The objectives of the study were to determine if workers potentially exposed to TiO_2 had higher risks of lung cancer, chronic respiratory disease, pleural thickening/plaques, or pulmonary fibrosis than referent groups.

Of the 2,477 male workers, 1,576 were potentially exposed to TiO_2. Other exposures included $TiCl_4$, pigmentary potassium titinate (PKT), and asbestos. (The $TiCl_4$-exposed workers were evaluated in Fayerweather et al. [1992]). Quantitative results from exposure monitoring or sampling performed after 1975 may have been included in the study; however, it was unclear what exposure measurements were available after 1975 and how they were used. Committees (not described) were established at the plants to estimate TiO_2 exposures for all jobs.

Table 2–1. Summary of epidemiologic studies of workers exposed to TiO_2.*

Reference and country	Study design, cohort, and follow-up	Subgroup	Number of deaths or cases in subgroup	Risk measure	95% CI	Adjusted for smoking	Comments
Boffetta et al. [2001], Canada	Population-based case-control study of 857 cases of histologically confirmed lung cancer diagnosed from 1979 to 1985 in men aged 35–70. Controls were randomly selected healthy residents (n = 533) and persons with cancers of other organs (n = 533).†	Ever exposed to TiO_2	33	OR = 0.9	0.5–1.5	Yes	TiO_2 exposures were estimated by industrial hygienists based on occupational histories collected by Siemiatycki et al. [1991] and other sources.
		Substantial exposure to TiO_2	8	OR = 1.0	0.3–2.7		"Substantial" exposure defined as exposure for >5 years at a medium or high frequency and concentration.
		Level of exposure: Low	25	OR = 0.9	0.5–1.7		
		Medium	6	OR = 1.0	0.3–3.3		
		High	2	OR = 0.3	0.07–1.9		
		Duration of exposure: 1–21 years	17	OR = 1.0	0.5–2.0		Lung cancer ORs were adjusted for age, family income, ethnicity, respondent (i.e., self or proxy), and smoking.
		≥ 22 years	16	OR = 0.8	0.4–1.6		Small number of cases ever exposed to TiO_2 (n = 33). Limitations include self- or proxy-reporting of occupational exposures.
		Exposed to TiO_2 fumes	5	OR = 9.1	0.7–118		Most TiO_2 fume-exposed cases (n = 5) and controls (n = 1) were also exposed to chromium and nickel.

(Continued)

See footnotes at end of table.

Table 2–1 (Continued). Summary of epidemiologic studies of workers exposed to TiO_2.*

Reference and country	Study design, cohort, and follow-up	Subgroup	Number of deaths or cases in subgroup	Risk measure	95% CI	Adjusted for smoking	Comments
Boffetta et al. [2004], Finland, France, Germany, Italy, Norway, United Kingdom	Retrospective cohort mortality study of 15,017 workers (14,331 men) employed ≥ 1 month in 11 TiO_2 production facilities and followed for mortality from 1950–1972 until 1997–2001 (follow-up period varied by country). Employment records were complete from 1927–1969 until 1995–2001.	Male lung cancer: Cumulative respirable TiO_2 dust exposure ($mg/m^3 \cdot year$):				Smoking data were available for 5,378 workers, but "since most available smoking data refer to recent years, no direct adjustment of risk estimates was attempted" [Boffetta et al. 2004].	No evidence of increased mortality risk with increasing cumulative TiO_2 dust exposure. (P-values for tests of linear trend were 0.5 and 0.6 for lung cancer mortality and nonmalignant respiratory disease mortality, respectively).
		0–0.73	53	RR = 1.00	Reference category		Estimated cumulative TiO_2 dust exposure was derived from job title and work history. Exposure indices were not calculated when > 25% of the occupational history or > 5 years were missing.
		0.73–3.43	53	RR = 1.19	0.80–1.77		
		3.44–13.19	52	RR = 1.03	0.69–1.55		
		13.20+	53	RR = 0.89	0.58–1.35		
		Male nonmalignant respiratory diseases: Cumulative respirable TiO_2 dust exposure ($mg/m^3 \cdot year$):					SMRs were not significantly increased for any cause of death except male lung cancer (SMR = 1.23; 95% CI 1.10–1.38; 306.5 deaths observed).
		0–0.8	40	RR = 1.00	Reference category		
		0.9–3.8	39	RR = 1.23	0.76–1.99		
		3.9–16.1	40	RR = 0.91	0.56–1.49		Female workers were not included in most statistical analyses because of small number of deaths (n = 33).
		16.2+	39	RR = 1.12	0.67–1.86		

(Continued)

See footnotes at end of table.

Table 2–1 (Continued). Summary of epidemiologic studies of workers exposed to TiO_2.*

Reference and country	Study design, cohort, and follow-up	Subgroup	Number of deaths or cases in subgroup	Risk measure	95% CI	Adjusted for smoking	Comments
Chen and Fayerweather [1988], United States	Mortality, morbidity, and nested case-control study of male, wage-grade employees of two TiO_2 production plants. Of 2,477 male employees, 1,576 were potentially exposed to TiO_2. Study subjects worked >1 year before January 1, 1984. Mortality was followed from 1935 through 1983 and compared with U.S. white male mortality rates or company rates. Cancer and chronic respiratory disease incidence cases from 1956–1985 were available from company insurance registry. Case-control methods were applied to findings from 398 chest X-ray films from current male employees as of January 1, 1984.	Lung cancer deaths 1935–1983	9	O/E = 0.52 (national rates)	11–24‡	Smoking histories were available for current workers; only use in X-ray case-control study was reported.	No statistically significant association or trends were reported. However, study has limitations (see text). Unclear source and exposure history of 898 controls in nested case-control study—may have been from company disease registry rather than entire worker population. Lung cancer OR was adjusted for age and exposure to $TiCl_4$, potassium titinate, and asbestos. "Chronic respiratory disease" was not defined. Controls (n = 372) for pleural thickening case-control study were active employees with normal chest X-ray findings. ORs were adjusted for age, current cigarette smoking habits, and exposure to known respiratory hazards (not defined).
		Lung cancer deaths 1957–1983	9	O/E = 0.59 (company rates)	9–22‡		
		Lung cancer cases 1956–1985	8	O/E = 1.04 (company rates)	3–13‡		
		Lung cancer cases (case-control study)	16	OR = 0.6	Not reported		
		Chronic respiratory disease cases (case-control study)	88	OR = 0.8	Not reported		
		Pleural thickening/plaque cases (case-control study)	22	OR = 1.4§	Not reported		

See footnotes at end of table.

(Continued)

Table 2–1 (Continued). Summary of epidemiologic studies of workers exposed to TiO_2.*

Reference and country	Study design, cohort, and follow-up	Subgroup	Number of deaths or cases in subgroup	Risk measure	95% CI	Adjusted for smoking	Comments
Fryzek [2004], Fryzek et al. [2003, 2004a], United States	Retrospective cohort mortality study of 409 female and 3,832 male workers employed ≥ 6 months on or after January 1, 1960, at four TiO_2 production facilities. The cohort was followed for mortality until the end of 2000. Mortality rates by sex, age, race, time period, and State where plant was located were used for numbers of expected deaths. Thirty-five percent (n = 1,496) of workers were employed in jobs with high potential TiO_2 dust exposure (i.e., packers, micronizers, and addbacks).	Trachea, bronchus, lung cancer deaths	61	SMR = 1.0	0.8–1.3	No	Lung cancer and nonmalignant respiratory disease SMRs not elevated significantly with exception of lung cancer death in the subgroup of shortest-term workers (0–9 years worked) with >20 years since first hire (SMR for cancer of trachea, bronchus, lung = 1.5; 95% CI 1.0–2.3; $P < 0.05$; number of deaths in subgroup not reported). Internal analyses with models found no significant exposure-response trends for those diseases or total cancers. Study limitations: (1) short follow-up period (avg. 21 years) and about half the cohort born after 1940; (2) more than half worked fewer than 10 years; (3) limited data on nonoccupational factors (e.g., smoking).
		High potential TiO_2 exposure jobs	11	SMR = 1.0	0.5–1.7		
		Nonmalignant respiratory disease deaths	31	SMR = 0.8	0.6–1.2		
		High potential TiO_2 exposure jobs	3	SMR = 0.4	0.1–1.3		
		All causes of death	533	SMR = 0.8	0.8–0.9		
		High potential TiO_2 exposure jobs	112	SMR = 0.7	0.6–0.9		

(Continued)

See footnotes at end of table.

Table 2–1 (Continued). Summary of epidemiologic studies of workers exposed to TiO$_2$.*

Reference and country	Study design, cohort, and follow-up	Subgroup	Number of deaths or cases in subgroup	Risk measure	95% CI	Adjusted for smoking	Comments
Fryzek [2004], Fryzek et al. [2003, 2004a], United States (Continued)							Study authors responded to a letter to the editor about modeling results, particularly the negative lung cancer exposure-response trend for cumulative TiO$_2$ exposure category and duration of exposure [Beaumont et al. 2004]. Response included additional analyses that yielded hazard ratios similar to original ones [Fryzek 2004; Fryzek et al. 2003, 2004a].
							914 full-shift or near full-shift personal air samples for TiO$_2$ dust were used in the analysis. Mean TiO$_2$ dust concentrations declined from 13.7 mg/m^3 ±17.9 (21 samples) in 1976–1980 to 3.1 mg/m^3 ± 6.1 (357 samples) in 1996–2000. They were 6.2 ± 9.4 mg/m^3 (686 samples) in jobs with high potential for TiO$_2$ exposure.
							(Continued)

See footnotes at end of table.

Table 2–1 (Continued). Summary of epidemiologic studies of workers exposed to TiO_2.*

Reference and country	Study design, cohort, and follow-up	Subgroup	Number of deaths or cases in subgroup	Risk measure	95% CI	Adjusted for smoking	Comments
Ramanakumar et al. [2008], Canada	Two population-based lung cancer case-control studies:	Pooled data from Studies I and II:				Yes	Study I similar to Boffetta et al. [2001] (above).
	Study I (1979–1986): 857 histologically confirmed lung cancer cases 1979 to 1985 in men aged 35–70. Controls were randomly selected healthy residents (n = 533) and 1,349 persons with cancers of other organs (n = 533).	Any TiO_2 exposure	76	OR = 1.0	0.8–1.5		TiO_2 exposures were estimated by industrial hygienists based on occupational histories collected by Siemiatycki et al. [1991] and other sources.
		Nonsubstantial TiO_2 exposure	68	OR = 1.0	0.6–1.7		"Substantial" exposure defined as exposure for >5 years at a medium or high frequency and concentration, excluding the 5 years before diagnosis or interview, with confidence of probable or definite occurrence.
		Substantial TiO_2 exposure	8	OR = 1.2	0.4–3.6		
	Study II (1995–2001): 765 histologically confirmed male and 471 female lung cancer cases with 899 male and 613 female population-based controls.						Lung cancer ORs were adjusted for age, family income, ethnicity, respondent (i.e., self or proxy), years of schooling, tobacco smoking (3 variables), and exposure to at least one of the other job hazards (i.e., cadmium compounds, asbestos, silica).
							Small number of cases in "substantial" TiO_2 exposure category (n = 8 males; 0 females).
							Occupational exposures reported by self- or surrogate respondents.

(Continued)

See footnotes at end of table.

Table 2–1 (Continued). Summary of epidemiologic studies of workers exposed to TiO$_2$.*

Reference and country	Study design, cohort, and follow-up	Subgroup	Number of deaths or cases in subgroup	Risk measure	95% CI	Adjusted for smoking	Comments
Siemiatycki et al. [1991], Canada	Population-based case-control study of 3,730 histologically confirmed cases of 20 types of cancer diagnosed from September 1979 to June 1985 in men aged 35–70. 140 cases had some occupational TiO$_2$ exposure. Two control groups were used: 533 population-based controls and a group of cancer patients.	Lung cancer cases with any occupational TiO$_2$ exposure	38	OR = 1.0	0.7–1.5**	Yes	Results provide little information about TiO$_2$-specific effects because this study evaluated 293 exposures, including TiO$_2$. Exposure was estimated by "chemist-hygienists" based on occupational histories. "Substantial" exposure defined as >10 years in the industry or occupation up to 5 years before onset [Siemiatycki et al. 1991, p. 122].
		Lung cancer cases with "substantial" occupational TiO$_2$ exposure	5	OR = 2.0	0.6–7.4**		
		Squamous cell lung cancer cases with any occupational TiO$_2$ exposure (population-based controls)	20	OR = 1.6	0.9–3.0**		
		Squamous cell lung cancer cases with "substantial" occupational TiO$_2$ exposure	2	OR = 1.3	0.2–9.8**		
		Bladder cancer cases with any occupational TiO$_2$ exposure (cancer patient controls)	28	OR = 1.7	1.1–2.6**		
		Substantial occupational TiO$_2$ exposure	3	OR = 4.5	0.9–22.0**		

*Abbreviations: CI = confidence interval; O/E = observed number of deaths or cases divided by expected number of deaths or cases; OR = odds ratio; RR = relative risk; SMR = standardized mortality ratio; TiO$_2$ = titanium dioxide.
†Number of controls in Boffetta et al. [2001] subgroups: 43 ever exposed, 9 substantial exposure; 29 low exposure; 9 medium exposure; 5 high exposure; 22 worked 1–21 years; 21 worked ≥ 22 years.
‡90% acceptance range for the expected number of deaths or cases
§Reported as "not statistically significantly elevated."
**90% CI.

A cumulative exposure index, duration of exposure, and TWA exposure were derived and used in the analyses (details not provided).

Chest radiographic examination was used to detect fibrosis and pleural abnormalities, and the most recent chest X-ray of active employees (on 1/1/1984) was read blindly by two B-readers. Chest X-ray films were not available for retired and terminated workers.

Observed numbers of cancer morbidity cases (i.e., incident cases) were compared to expected numbers based on company rates. Observed numbers of deaths were compared to expected numbers from company rates and national rates. Ninety percent (90%) "acceptance ranges" were calculated for the expected numbers of cases or deaths. The nested case-control study investigated decedent lung cancer and chronic respiratory disease, incident lung cancer and chronic respiratory disease (not described), and radiographic chest abnormalities. Incidence data from the company's insurance registry were available from 1956 to 1985 for cancer and chronic respiratory disease. Mortality data from 1957 to 1983 were obtained from the company mortality registry. The study reported the number of observed deaths for the period 1935–1983; the source for deaths prior to 1957 is not clear. Vital status was determined for "about 94%" of the study cohort and death certificates were obtained for "about 94%" of workers known to be deceased.

The observed number of deaths from all cancers was lower than the expected number based on U.S. mortality rates; however, the observed number of deaths from all causes was greater than the expected number when based on company mortality rates (194 deaths observed; 175.5 expected; 90% acceptance range for the expected number of deaths = 154–198). Lung cancer deaths were lower than the expected number based on national rates (9 deaths observed/17.3 expected = 0.52; 90% acceptance range for the expected number of deaths = 11–24) and company rates (9 deaths observed/15.3 deaths expected = 0.59; 90% acceptance range for the expected number of deaths = 9–22). Lung cancer morbidity was not greater than expected (company rates; 8 cases observed; 7.7 expected; 90% acceptance range for the expected number of cases = 3–13).

Nested case-control analyses found no association between TiO_2 exposure and lung cancer morbidity after adjusting for age and exposure to $TiCl_4$, PKT, and asbestos (16 lung cancer cases; 898 controls; TiO_2 odds ratio [OR] = 0.6). The OR did not increase with increasing average exposure, duration of exposure, or cumulative exposure index. No statistically significant positive relationships were found between TiO_2 exposure and cases of chronic respiratory disease (88 cases; 898 noncancer, nonrespiratory disease controls; TiO_2 OR = 0.8). Chest X-ray findings from 398 films showed few abnormalities—there were four subjects with "questionable nodules" but none with fibrosis. Pleural thickening or plaques were present in 5.6% (n = 19) of the workers potentially exposed to TiO_2 compared with 4.8% (n = 3) in the unexposed group. Case-control analyses of 22 cases and 372 controls with pleural abnormalities found a nonstatistically significant OR of 1.4 for those potentially exposed and no consistent exposure-response relationship.

This study did not report statistically significant increased mortality from lung cancer, chronic respiratory disease, or fibrosis associated with titanium exposure. However, it has limitations: (note: the study component or information affected by the limitation is mentioned, when possible) (1) Existence of quantitative exposure data for respirable TiO_2 after 1975 is implied; the type of measurement (e.g., total, respirable, or submicrometer), type of sample (e.g., area or personal), number of samples, sampling location and times, nature of samples (e.g., epidemiologic study or compliance survey), and breathing zone particle sizes were not reported. (Exposure data were used in the nested case-control analyses of morbidity and mortality.) (2) The report did not describe the number of workers, cases, or deaths in each exposure duration quartile which could contribute to understanding of all component results. (3) The presence of other chemicals and asbestos could have acted as confounders. (4) Incidence and mortality data were not described in detail and could have been affected by the healthy worker effect. (5) Company registries were the only apparent source for some incidence and mortality information (e.g., company records may have been based on those workers eligible for pensions and thus not typical of the general workforce).

2.2.2 Fryzek et al. [2003]

Fryzek et al. [2003] conducted a retrospective cohort mortality study of 4,241 workers with potential exposure to TiO_2 employed on or after 1/1/1960 for at least 6 months at four TiO_2 production plants in the United States.

The plants used either a sulfate process or a chloride process to produce TiO_2 from the original ore. Nearly 2,400 records of air sampling measurements of sulfuric acid mist, sulfur dioxide, hydrogen sulfide, hydrogen chloride, chlorine, $TiCl_4$, and TiO_2 were obtained from the four plants. Most were area samples and many were of short duration. Full-shift or near full-shift personal samples (n = 914; time-weighted averaging not reported) for total TiO_2 dust were used to estimate relative exposure concentrations between jobs over time. Total mean TiO_2 dust levels declined from 13.7 mg/m³ in 1976–1980 to 3.1 mg/m³ during 1996–2000. Packers, micronizers, and addbacks had about 3 to 6 times higher exposure concentrations than other jobs. Exposure categories, defined by plant, job title, and calendar years in the job, were created to examine mortality patterns in those jobs where the potential for TiO_2 exposure was greatest.

Mortality of 409 female workers and 3,832 male workers was followed until 12/31/2000 (average follow-up time = 21 years; standard deviation = 11 years). The number of expected deaths was based on mortality rates by sex, age, race, time period, and the state where the plant was located, and standardized mortality ratios (SMRs) and confidence intervals (CIs) were calculated. Cox proportional hazards (PH) models, which adjusted for effects of age, sex, geographic area, and date of hire, were used to estimate relative risks (RR) of TiO_2 exposure (i.e., average intensity, duration, and cumulative exposure) in medium or high exposure groups versus the lowest exposure group.

Of the 4,241 workers (58% white, 22% non-white, 20% unknown race, 90% male), 958 did not have adequate work history information and were omitted from some plant analyses. Some company records from the early period may have been lost or destroyed; however, the study authors "found no evidence to support such an assumption" [Fryzek et al. 2003]. Wage status (i.e., hourly or salaried) was not described. Thirty-five percent of workers had been employed in jobs with the highest potential for TiO_2 exposure. Workers experienced a significantly low overall mortality (533 deaths, SMR = 0.8, 95% CI = 0.8–0.9; $P < 0.05$). No significantly increased SMRs were found for any specific cause of death and there were no trends with exposure. (Results were not reported by category of race). The number of deaths from trachea, bronchus, or lung cancer was not greater than expected (i.e., 61 deaths; SMR = 1.0; 95% CI = 0.8–1.3). However, there was a significant 50% elevated SMR for workers employed 9 years or less with at least 20 years since hire (SMR = 1.5 [cancer of trachea, bronchus, or lung]; 95% CI = 1.0–2.3; number of deaths not reported). SMRs for this cancer did not increase with increasing TiO_2 concentrations (i.e., as evidenced by job category in Table 6 of the study). Workers in jobs with greatest TiO_2 exposure had significantly fewer than expected total deaths (112 deaths; SMR = 0.7; 95% CI = 0.6–0.9) and mortality from cancers of trachea, bronchus, or lung was not greater than expected (11 deaths; SMR = 1.0; 95% CI = 0.5–1.7). Internal analyses (i.e., Cox PH models) revealed no significant trends or exposure-response associations for total cancers, lung cancer, or other causes of death. No association between TiO_2 exposure and increased risk of cancer death was observed in this study (i.e., Fryzek et al. [2003]).

Limitations of this study include (1) about half the cohort was born after 1940, lung cancer in these younger people would be less frequent and the latency from first exposure to TiO_2 would be short, (2) duration of employment was often quite short, (3) no information about ultrafine exposures (probably because collection methods were not available over the course of the study), and (4) limited data on nonoccupational factors (e.g., smoking). Smoking information abstracted from medical records from 1960 forward of 2,503 workers from the four plants showed no imbalance across job groups. In all job groups, the prevalence of smoking was about 55% and it declined over time by decade of hire. However, the information was inadequate for individual adjustments for smoking [Fryzek et al. 2003].

Fryzek et al. [Fryzek 2004; Fryzek et al. 2004a] performed additional analyses in response to a suggestion that the RRs may have been artificially low, especially in the highest category of cumulative exposure, because of the statistical methods used [Beaumont et al. 2004]. These analyses yielded hazard ratios similar to those in the original analysis and found no significant exposure-response relationships for lung cancer mortality and cumulative TiO_2 exposure (i.e., "low," "medium," "high") with either a time-independent exposure variable or a time-dependent exposure variable and a 15-year exposure lag (adjusted for age, sex, geographic area, and date of hire) [Fryzek 2004; Fryzek et al. 2004a]. The hazard ratio for trachea, bronchus, and lung cancer from "medium" cumulative TiO_2 exposure (15-year lag) was greater than 1.0 (hazard ratio for "medium" cumulative exposure, time-dependent exposure variable and 15-year lag = 1.3; 95% CI = 0.6–2.8) and less than 1.0 for "high": (hazard ratio = 0.7; 95% CI = 0.2–1.8; "low" was the referent group) [Fryzek 2004; Fryzek et al. 2004a].

2.2.3 Boffetta et al. [2001]

Boffetta et al. [2001] reevaluated lung cancer risk from exposure to TiO_2 in a subset of a population-based case-control study of 293 substances including TiO_2 (i.e., Siemiatycki et al. [1991]; see Table 2–1 for description of Siemiatycki et al. [1991]).

Histologically confirmed lung cancer cases (n = 857) from hospitals and noncancer referents were randomly selected from the population of Montreal, Canada. Cases were male, aged 35 to 70, diagnosed from 1979 to 1985, and controls were 533 randomly selected healthy residents and 533 persons with cancer in other organs.

Job information was translated into a list of potential exposures, including all Ti compounds and TiO$_2$ as dust, mist, or fumes. Using professional judgment, industrial hygienists assigned qualitative exposure estimates to industry and job combinations worked by study subjects, based on information provided in interviews with subjects, proxies, and trained interviewers and recorded on a detailed questionnaire. The exposure assessment was conducted blindly (i.e., case or referent status not known). Duration, likelihood (possible, probable, definite), frequency (<5%, 5%–30%, >30%), and extent (low, medium, high) of exposure were assessed. Those with probable or definite exposure for at least 5 years before the interview were classified as "exposed." Boffetta et al. [2001] classified exposure as "substantial" if it occurred for more than 5 years at a medium or high frequency and level. (Siemiatycki et al. [1991] used a different definition and included five workers exposed to titanium slag who were excluded by Boffetta et al. [2001]; see Table 2–1). Only 33 cases and 43 controls were classified as ever exposed to TiO$_2$ (OR = 0.9; 95% CI = 0.5–1.5). Results of unconditional logistic models were adjusted for age, socioeconomic status, ethnicity, respondent status (i.e., self or proxy), tobacco smoking, asbestos, and benzo(a)pyrene (BAP) exposure. No trend was apparent for estimated frequency, level, or duration of exposure. The OR was 1.0 (95% CI = 0.3–2.7) for medium or high exposure for at least 5 years. Results did not depend on choice of referent group, and no significant associations were found with TiO$_2$ exposure and histologic type of lung cancer. The likelihood of finding a small increase in lung cancer risk was limited by the small number of cases assessed. However, the study did find an excess risk for lung cancer associated with both asbestos and BAP, indicating that the study was able to detect risks associated with potent carcinogens. The study had a power of 86% to detect an OR of 2 at the 5% level, and 65% power for an OR of 1.5.

Limitations of this study include (1) self-reporting or proxy reporting of exposure information; (2) use of surrogate indices for exposure; (3) absence of particle size characterization; and (4) the nonstatistically significant lung cancer OR for exposure to TiO$_2$ *fumes*, which was based on a small group of subjects and most were also exposed to nickel and chromium (5 cases; 1 referent; OR = 9.1; 95% CI = 0.7–118). In addition, exposures were limited mainly to those processes, jobs, and industries in the Montreal area. For example, the study probably included few, if any, workers who manufactured TiO$_2$. Most workers classified as TiO$_2$-exposed were painters and motor vehicle mechanics and repairers with painting experience; the highly exposed cases mixed raw materials for the manufacture of TiO$_2$-containing paints and plastics.

2.2.4 Boffetta et al. [2004]

Boffetta et al. [2004] conducted a retrospective cohort mortality study of lung cancer in 15,017 workers (14,331 men, 686 women) employed at least 1 month in 11 TiO$_2$ production facilities in six European countries. The factories produced mainly pigment-grade TiO$_2$. Estimated cumulative occupational exposure to respirable TiO$_2$ dust was derived from job title and work history. Observed numbers of deaths were compared with expected numbers based on national rates; exposure-response relationships within the cohort were evaluated using the Cox PH model. Few deaths occurred in female workers (n = 33); therefore, most analyses did not include female deaths. The follow-up period ranged from 1950–1972 until 1997–2001; 2,619 male and 33 female workers were reported as deceased. (The follow-up periods probably have a range of years because the follow-up procedures varied with the participating countries.) The cause of death was not known for 5.9% of deceased cohort members. Male lung cancer was the only cause of death with a statistically significant SMR (SMR = 1.23; 95% CI = 1.10–1.38; 306.5 deaths [not a whole number because of correction factors for missing deaths]). However, the Cox regression analysis of male lung cancer mortality found no evidence of increased risk with increasing cumulative respirable TiO_2 dust exposure (P-value for test of linear trend = 0.5). There was no consistent and monotonic increase in SMRs with duration of employment, although workers with more than 15 years of employment had slightly higher SMRs than workers with 5 to 15 years of employment and an effect of time since first employment was suggested for workers employed more than 10 years. (The authors indicated that the increase in lung cancer mortality with increasing time since first employment could be "explained by the large contribution of person-years to the categories with longest time since first employment from countries such as Germany, with increased overall lung cancer mortality" [Boffetta et al. 2004].) For male nonmalignant respiratory disease mortality, the number of observed deaths was lower than the expected number (SMR = 0.88; 95% CI = 0.77–1.02; 201.9 deaths observed; 228.4 expected), and there was no evidence of an exposure-response relationship.

The authors suggested that lack of exposure-response relationships may have been related to a lack of (1) statistical power or (2) inclusion of workers who were employed before the beginning of the follow-up period when exposure concentrations tended to be high. (Regarding the latter point, the authors stated that this phenomenon could have occurred

and resulted in a loss of power, but "the results of the analysis on the inception cohort, composed of workers whose employment is entirely covered by the follow-up, are remarkably similar to the results of the whole cohort, arguing against survival bias" [Boffetta et al. 2004]). The authors also suggested that the statistically significant SMR for male lung cancer could represent (1) heterogeneity by country, which the authors thought should be explained by chance and differences in effects of confounders (see next item), rather than factors of TiO_2 dust exposure; (2) differences in the effects of potential confounders, such as smoking or occupational exposure to lung carcinogens; or (3) use of national reference rates instead of local rates.

2.2.5 Ramanakumar et al. [2008]

Ramanakumar et al. [2008] analyzed data from two large, population-based case-control studies conducted in Montreal, Canada, and focused on lung cancer risk from occupational exposure to four agents selected from a large set of agents and mixtures: carbon black, TiO_2, industrial talc, and cosmetic talc. Results from the first study (i.e., Study I by Boffetta et al. [2001]) involving TiO_2 were published and are described above (see Section 2.2.3). Subject interviews for Study I were conducted from 1982–1986 and included 857 lung cancer cases in men aged 35–70 years. Study II's interviews were conducted from 1995–2001 and included men and women aged 35–75 years (765 male lung cancer cases, 471 female lung cancer cases). In both studies, lung cancer cases were obtained from 18 of the largest hospitals in the metropolitan Montreal area and confirmed histologically. Controls were randomly sampled from the population. Study I had an additional control group of persons with non-lung cancers (Study I: 533 population controls and 1,349 cancer controls; Study II: 899 male controls and 613 female controls). Exposure assessment methods were similar to those described in Boffetta et al. [2001] and Siemiatycki et al. [1991] and included estimates by industrial hygienists based on job histories (see Section 2.2.3 and Table 2–1). Major occupations of cases and controls with TiO$_2$ exposure (n = 206) were painting, paper hanging, and related occupations (37%); construction laborer, grinder chipper and related occupations; motor-body repairmen; and paint plant laborer. Unconditional multivariate logistic regression models estimated the association between the exposure and lung cancer and included potential confounders of age, ethnicity, education, income, type of respondent (i.e., self or surrogate), smoking history, and exposure to at least one other known occupational hazard (i.e., cadmium compounds, silica, or asbestos). Association of substance with cell type (i.e., squamous cell, adenocarcinoma, small cell) was also assessed. ORs for lung cancer and exposure to TiO$_2$ (i.e., "any," "nonsubstantial," or "substantial") were not statistically significantly increased, exposure-response trends were not apparent, and there was no evidence of a confounding effect of smoking or other confounder or an association with histologic type (results by histologic type were not shown). In the pooled analysis of cases, controls, and sexes from both studies, the ORs were 1.0 (95% CI = 0.8–1.5), 1.0 (95% CI = 0.6–1.7), and 1.2 (95% CI = 0.4–3.6) for TiO$_2$ exposure categories of "any," "nonsubstantial," and "substantial," respectively. ORs were adjusted for the possible confounders mentioned above [Ramanakumar et al. 2008].

Limitations of this study are similar to those for the Boffetta et al. [2001] study and include (1) self-reporting or proxy reporting of exposure information, (2) use of surrogate indices for exposure, (3) absence of particle size characterization, (4) few lung cancer cases in the "substantial exposure" to TiO$_2$ category (n = 8, both studies combined) and no female cases in that category, and (5) exposures limited mainly to those processes, jobs, and industries in the Montreal area.

2.3 Summary of Epidemiologic Studies

In general, the five epidemiologic studies of TiO$_2$-exposed workers represent a range of environments, from industry to population-based, and appear to be reasonably representative of worker exposures over several decades. One major deficiency is the absence of any cohort studies of workers who handle or use TiO$_2$ (rather than production workers).

Overall, these studies provide no clear evidence of elevated risks of lung cancer mortality or morbidity among those workers exposed to TiO$_2$ dust.

Nonmalignant respiratory disease mortality was not increased significantly (i.e., $P < 0.05$) in any of the three epidemiologic studies that investigated it. Two of the three retrospective cohort mortality studies found small numbers of deaths from respiratory diseases other than lung cancer, and the number of pneumoconiosis deaths within that category was not reported, indicating that these studies may have lacked the statistical power to detect an increased risk of mortality from TiO$_2$-associated pneumoconiosis (i.e., Chen and Fayerweather [1988]: 11 deaths from nonmalignant diseases of the respiratory system; Fryzek et al. [2003]: 31 nonmalignant respiratory disease deaths). The third study had a larger number of male deaths from nonmalignant respiratory disease and found no excess mortality (SMR = 0.88; 201.9 deaths observed; 95% CI = 0.77–1.02) [Boffetta et al. 2004]. None of the studies reported an SMR for pneumoconiosis mortality; Boffetta et al. [2004] did discuss four pleural cancer deaths, although the number of observed deaths was lower than expected number, based on national rates. Boffetta et al. [2004] suggested that "mortality data might not be very sensitive to assess risks of chronic respiratory diseases."

In addition to the methodologic and epidemiologic limitations of the studies, they were not designed to investigate the relationship between TiO_2 particle size and lung cancer risk, an important question for assessing the potential occupational carcinogenicity of TiO_2. Further research is needed to determine whether such epidemiologic studies of TiO_2-exposed workers can be designed and conducted and to also study workers who manufacture or use products that contain TiO_2 (see Chapter 7 Research Needs).

3 Experimental Studies in Animals and Comparison to Humans

3.1 In Vitro Studies

3.1.1 Genotoxicity and Mutagenicity

Titanium dioxide (TiO$_2$) did not show genotoxic activity in several standard assays: cell-killing in deoxyribonucleic acid (DNA)-repair deficient *Bacillus subtilis*, mutagenesis in *Salmonella typhimurium* or *E. coli*, or transformation of Syrian hamster embryo cells (particle size and crystal form not specified) [IARC 1989]. TiO$_2$ was not genotoxic in a Drosphila wing spot test [Tripathy et al. 1990] or mutagenic in mouse lymphoma cells [Myhr and Caspary 1991] (particle size and crystal structure not provided in either study). More recent genotoxicity studies have shown that TiO$_2$ induced chromosomal changes, including micronuclei in Chinese hamster ovary cells (particularly when a cytokinesis-block technique was employed) [Lu et al. 1998] and sister chromatid exchanges and micronuclei in lymphocytes [Türkez and Geyikoğlu 2007] (particle size and crystal form not provided in either study). Photo-illumination of TiO$_2$ (anatase/rutile samples of various ratios; particle size not known) catalyzed oxidative DNA damage in cultured human fibroblast cells, which the assay indicated was due to hydroxyl radicals [Dunford et al. 1997]. Ultrafine TiO$_2$ (particle diameter < 100 nm; crystal structure not provided) induced apoptosis in Syrian hamster embryo cells [Rahman et al. 2002] and in cultured human lymphoblastoid cells [Wang et al. 2007a]. Sanderson et al. [2007] provided additional physical-chemical data on the ultrafine TiO$_2$ material studied in Wang et al. [2007a] (anatase ≥ 99%; mean particle diameter = 6.57; surface area = 147.9 m^2/g). DNA damage (micronuclei) was produced in human lymphoblastoid cells at a 65 µg/ml dose without excessive cell killing (20%) [Sanderson et al. 2007; Wang et al. 2007a].

Ultrafine TiO$_2$ (80% anatase, 20% rutile) was not genotoxic without UV/vis light irradiation in treated cells but did show dose-dependent increase in chromosome aberrations in a Chinese hamster cell line with photoactivation [Nakagawa et al. 1997]. Greater photocatalytic activity by mass was observed for anatase than for rutile TiO$_2$ (specific surface areas: 14.6 and 7.8 m^2/g, respectively) [Kakinoki et al. 2004] and for anatase-rutile mixtures (e.g., 80% anatase, 20% rutile [Behnajady et al. 2008]) compared to either particle type alone (0.5% wt anatase due to photoinduced interfacial electron transfer from anatase to rutile [Kawahara et al. 2003]). Particle size influenced oxidative DNA damage in cultured human bronchial epithelial cells, which was detected for 10- and 20-nm but not 200-nm diameter anatase TiO$_2$ [Gurr et al. 2005]. In the absence of photoactivation, an anatase-rutile mixture (50%, 50%; diameter = 200 nm) induced higher oxidative DNA damage than did the pure anatase or rutile (200 nm each) [Gurr et al. 2005]. Overall, these studies indicate that TiO$_2$ exhibits genotoxicity (DNA damage) under certain conditions but not mutagenicity (genetic alteration) in the assays used.

3.1.2 Oxidant Generation and Cytotoxicity

TiO_2 is considered to be of relatively low inherent toxicity, although the crystal phase can influence the particle surface properties and cytotoxicity *in vitro*. Sayes et al. [2006] reported that nano-anatase produced more ROS and was more cytotoxic than nano-rutile, but only after UV irradiation. ROS generation by cells *in vitro* (mouse BV2 microglia) treated with P25 ultrafine TiO_2 (70% anatase and 30% rutile) was suggested as the mechanism for damaging neurons in complex grain cell cultures [Long et al. 2007]. In contrast, Xia et al. [2006] reported that TiO_2 (80% anatase, 20% rutile; ~25 nm diameter) did not induce oxidative stress or increase the heme oxygenase 1 (HO-1) expression in phagocytic cells (RAW 264.7), which the authors suggested may be due to passivation of the particle surfaces by culture medium components or neutralization by available antioxidants. In a study comparing *in vitro* cellular responses to P25 ultrafine TiO_2 (21 nm particle size) and fine TiO_2 (1 μm particle size) at exposure concentrations of 0.5–200 μg/ml, the generation of ROS was significantly elevated relative to controls after 4-hr exposure to either fine or ultrafine TiO_2, although the ROS induced by ultrafine TiO_2 was greater than that of fine TiO_2 at each exposure concentration; no cytotoxicity was observed 24 hours after treatment at any of these doses [Kang et al. 2008]. Thus, photoactivation appears to be an important mechanism for increasing cytotoxicity of TiO_2, especially for formulations containing anatase. However, TiO_2 cytotoxicity is low relative to more inherently cytotoxic particles such as crystalline silica [Duffin et al. 2007].

3.1.3 Effects on Phagocytosis

Renwick et al. [2001] reported that both fine and ultrafine TiO_2 particles (250 and 29 nm mean diameter; 6.6 and 50 m^2/g surface area, respectively) reduced the ability of J774.2 mouse alveolar macrophages to phagocytose 2 μm latex beads at doses of 0.39 and 0.78 μg/mm^2. Ultrafine particles (TiO_2 and carbon black) impaired macrophage phagocytosis at lower mass doses than did their fine particle counterparts, although this effect was primarily seen with ultrafine carbon black (254 m^2/g surface area).

Möller et al. [2002] found that ultrafine TiO_2 (20 nm diameter), but not fine TiO_2 (220 nm diameter), significantly retarded relaxation and increased cytoskeletal stiffness in mouse alveolar macrophages (J774A.1 cell line) at a dose of 320 μg/ml. Ultrafine TiO_2 inhibited proliferation in J774A.1 cells to a greater extent than did fine TiO_2 (100 μg/ml dose; 50% and 90% of control proliferation, respectively). In primary alveolar macrophages (BD-AM, isolated from beagle dogs by bronchoalveolar lavage [BAL*]), either fine or ultrafine TiO_2 (100 μg/ml dose) caused moderate retardation of relaxation. Neither fine nor ultrafine TiO_2 caused impaired phagocytosis of latex microspheres in BD-AM cells, but both particle sizes significantly impaired phagocytosis in J774A.1 cells (~65% of control level by fine or ultrafine TiO_2). Ultrafine TiO_2 reduced the fraction of viable cells (either J774A.1 or BD-AM) to a greater extent than did fine TiO_2 (at 100 μg/ml dose).

These *in vitro* studies provide mechanistic information about how particle-macrophage interactions may influence cell function and disease processes *in vivo*. Overall, ultrafine TiO_2 impairs alveolar macrophage function to a greater extent than does fine TiO_2, which

---

*Bronchoalveolar lavage (BAL) is a procedure for washing the lungs to obtain BAL fluid (BALF), which contains cellular and biochemical indicators of lung health and disease status.

may also relate to the greater inflammatory response to ultrafine TiO$_2$ at a given mass dose.

3.2 In Vivo Studies in Rodent Lungs

3.2.1 Intratracheal Instillation

3.2.1.1 Short-term follow-up

Studies with male Fischer 344 rats instilled with 500 μg of TiO$_2$ of four different particle sizes and two crystal structures (rutile: 12 and 230 nm; anatase: 21 and 250 nm) indicate that the ultrafine TiO$_2$ particles (~20 nm) are translocated to the lung interstitium (interstitialized) to a greater extent and cleared from the lungs more slowly than fine TiO$_2$ particles (~250 nm) [Ferin et al. 1992]. Other intratracheal instillation (IT) studies conducted by the same laboratory showed that ultrafine TiO$_2$ particles produced a greater pulmonary inflammation response than an equal mass dose of fine TiO$_2$ particles. The greater toxicity of the ultrafine particles was related to their larger particle surface area dose and their increased interstitialization [Oberdörster et al. 1992].

In a study of four types of ultrafine particles (TiO$_2$, nickel, carbon black, and cobalt; particle sizes 14–20 nm), male Wistar rats (aged 10–12 weeks) were administered IT doses of 125 μg of particles, and BAL was performed after 4 or 18 hours. Ultrafine TiO$_2$ was the least toxic and inflammogenic, although it was still significantly greater than the saline-treated controls [Dick et al. 2003]. The adverse pulmonary responses were associated with the free radical activity of the particles, which was very low for ultrafine TiO$_2$.

Höhr et al. [2002] observed that, for the same surface area, the inflammatory response (as measured by BAL fluid (BALF) markers of inflammation) in female Wistar rats to uncoated TiO$_2$ particles covered with surface hydroxyl groups (hydrophilic surface) was similar to that of TiO$_2$ particles with surface OCH$_3$-groups (hydrophobic surface) replacing OH-groups. The IT doses in this study were 1 or 6 mg fine (180 nm particle size; 10 m^2/g specific surface area) or ultrafine (20–30 nm; 40 m^2/g specific surface area) TiO$_2$, and BAL was performed 16 hours after the IT.

Ultrafine TiO$_2$ was more damaging than fine TiO$_2$ in the lungs of male Wistar rats [Renwick et al. 2004]. The particle mean diameters of the ultrafine and fine TiO$_2$ were 29 and 250 nm, respectively, and the specific surface areas were 50 and 6.6 m^2/g; the crystal structure was not specified [Renwick et al. 2004]. Twenty-four hours after instillation of ultrafine or fine TiO$_2$, rats treated with ultrafine TiO$_2$, but not fine TiO$_2$, had BALF elevations in the percentage of neutrophils (indicating inflammation), γ-glutamyl transpeptidase concentration (a measure of cell damage), protein concentration (a measure of epithelium permeability), and lactate dehydrogenase (LDH) (an indicator of cytotoxicity and cell death) [Renwick et al. 2004]. The 125 μg IT dose of either ultrafine or fine TiO$_2$ did not cause any significant adverse lung response in 24 hours. At 500 μg, the phagocytic ability of the alveolar macrophages was significantly reduced by exposure to particles of either size. The 500 μg IT dose of ultrafine TiO$_2$, but not fine TiO$_2$, was associated with an increased sensitivity of alveolar macrophages to a chemotactic stimulus, an effect which can reduce the macrophage mobility and clearance of particles from the lungs.

In a study that included both inhalation (see Section 3.2.3) and IT exposures to six different formulations of fine rutile TiO$_2$ (including uncoated or alumina- or amorphous silica-coated; particle size 290–440 nm; specific surface area

6–28 m^2/g), male 8-week old Sprague Dawley rats were dosed with 2 or 10 mg/kg by IT. Pulmonary inflammation (measured by polymorphonuclear leukocytes [PMNs] in BALF) was statistically significantly elevated at 24 hours in the rats administered 10 mg/kg for each of the coated or uncoated formulations. The coated TiO_2 formulations produced higher inflammation than the uncoated TiO_2 [Warheit et al. 2005].

In a study of rutile TiO_2 nanorods, inflammation responses were examined in BALF and whole blood in Wistar rats 24 hours after an IT dose of 1 or 5 mg/kg [Nemmar et al. 2008]. At both doses, the neutrophilic inflammation in BALF was significantly greater than the vehicle controls. The number of monocytes and granulocytes in blood was dose-dependently elevated, while the platelets were significantly reduced at the higher dose, indicating platelet aggregation.

Mice instilled with 1 mg fine TiO_2 (250 nm mean diameter) showed no evidence of inflammation at 4, 24, or 72 hrs after instillation as assessed by inflammatory cells in BALF and expression of a variety of inflammatory cytokines in lung tissue [Hubbard et al. 2002].

Adult male ICR mice (2 months old, 30 g) were exposed to ultrafine (nanoscale) TiO_2 (rutile, 21 nm average particle size; specific surface area of 50 m^2/g) or fine (microscale) TiO_2 (180–250 nm diameter; specific surface area of 6.5 m^2/g) by a single IT dose of either 0.1 or 0.5 mg per mouse [Chen et al. 2006]. One week later, the lungs showed "significant changes in morphology and histology" in the mice receiving the 0.1-mg dose of nanoscale TiO_2, including disruption of the alveolar septa and alveolar enlargement (indicating emphysema), type II pneumocyte proliferation, increased alveolar epithelial thickness, and accumulation of particle-laden alveolar macrophages. Nanoscale TiO_2 elicited a significantly greater increase in chemokines associated with pulmonary emphysema and alveolar epithelial cell apoptosis than did the microscale TiO_2.

A dose-response relationship was not seen, as the adverse effects of the 0.1-mg dose of nanoscale TiO_2 exceeded those of the 0.5-mg dose. "No significant pathological changes" were observed at either dose of the microscale TiO_2.

In summary, these short-term studies show that while TiO_2 was less toxic than several other particle types tested, TiO_2 did elicit pulmonary inflammation and cell damage at sufficiently high surface area doses (i.e., greater response to ultrafine TiO_2 at a given-mass dose).

3.2.1.2 Intermediate-term follow-up

Rehn et al. [2003] observed an acute (3-day) inflammatory response to instillation of ultrafine TiO_2 and found that the response from a single instillation decreased over time, returning to control levels by 90 days after the instillation. The reversibility of the inflammatory response to ultrafine TiO_2 contrasted with the progressive increase in inflammation over 90 days that was seen with crystalline silica (quartz) in the same study. This study also compared a silanized hydrophobic preparation of ultrafine TiO_2 to an untreated hydrophilic form and concluded that alteration of surface properties by silanization does not greatly alter the biological response of the lung to ultrafine TiO_2.

Three recent studies of various types of nanoscale or microscale TiO_2 used a similar experimental design, which involved IT dosing of male Crl:CD®(SD):IGS BR rats (approximately 8 weeks of age; 240–255 g body weight). Instilled particle doses were either 1 or 5 mg/kg, and BAL was performed at 24 hours, 1 week, 1 month, and 3 months after instillation [Warheit et al. 2006a,b; 2007]. Cell proliferation assays and histopathological examination were also performed. Min-U-Sil quartz was used as a positive control, and phosphate buffered saline (PBS) was the instillation vehicle in controls.

In the first study of two hydrophilic types of TiO_2 ("R-100" or "Pigment A"), rats were administered IT doses of either 1 or 5 mg/kg of either type of TiO_2, carbonyl iron, or Min-U-Sil quartz. Primary average particle sizes were 300 nm, 290 nm, ~1.2 μm, or ~1.5 μm, respectively [Warheit et al. 2006a]. Significantly elevated PMNs in BALF was observed for the two types of TiO_2 or carbonyl iron at 24 hours postexposure, but not at the later time points.

The second study compared two types of nanoscale TiO_2 rods (anatase, 92–233 nm length, 20–35 nm width; 26.5 m^2/g specific surface area), nanoscale TiO_2 dots (anatase, 5.8–6.1 nm spheres; 169 m^2/g specific surface area), and microscale rutile TiO_2 (300 nm primary particle diameter; 6 m^2/g specific surface area) [Warheit et al. 2006b]. A statistically significant increase in the percentage of PMNs in BALF was seen at the 5 mg/kg dose for all three TiO_2 materials tested (which was higher in the rats administered the nanoscale TiO_2) but returned to control levels at the 1-week time point. There were no statistically significant lung responses (inflammation or histopathology) to either the fine or the ultrafine TiO_2 at either dose (1 or 5 mg/kg) compared to controls at the 1-week to 3-month time points.

In the third study, comparisons were made of the lung inflammation, cytotoxicity, cell proliferation, and histopathological responses of two types of ultrafine rutile TiO_2, a fine rutile TiO_2, an ultrafine anatase-rutile mixture (80%, 20%) TiO_2, and Min-U-Sil quartz particles [Warheit et al. 2007]. Although the surface area of these particles varied from 5.8 to 53 m^2/g, the median particle sizes in the PBS instillation vehicle were similar (2.1–2.7 μm), perhaps due to agglomeration. The pulmonary inflammation (% PMNs) of the 5 mg/kg-dose group of anatase/rutile TiO_2 was statistically significantly greater than the PBS controls 24 hrs and 1 wk after IT (but not at 1 or 3 months), while the ultrafine and fine rutile TiO_2 groups did not differ significantly from controls at any time point. The tracheobronchial epithelial cell proliferation (% proliferating cells) at the 5 mg/kg-dose group of anatase/rutile TiO_2 was also statistically significantly greater than controls 24 hrs after IT (but not at the later time points), while the ultrafine and fine rutile TiO_2 groups did not differ significantly from controls at any time point. The two ultrafine rutile TiO_2 preparations were passivated with amorphous silica and alumina coatings to reduce their chemical and photoreactivity to a low level similar to that of the fine rutile TiO_2, while the ultrafine anatase/rutile TiO_2 was not passivated and was also more acidic. The ultrafine anatase/rutile TiO_2 was more chemically reactive in a Vitamin C assay measuring oxidation potential. These results suggest that the crystal phase and surface properties can influence the lung responses to TiO_2. In both studies, the Min-U-Sil quartz-instilled rats showed the expected persistent inflammation.

3.2.1.3 Long-term follow-up

In a study of the role of lung phagocytic cells on oxidant-derived mutations, rats were dosed by IT with either 10 or 100 mg/kg of fine TiO_2 (anatase; median diameter: 180 nm; surface area: 8.8 m^2/g) and held for 15 months [Driscoll et al. 1997]. Type II cells isolated from the rats dosed with 100 mg/kg fine TiO_2 exhibited an increased hypoxanthine-guanine phosphoribosyl transferase (*hprt*) mutation frequency, but type II cells isolated from rats treated with 10 mg/kg fine TiO_2 did not. Neutrophil counts were significantly elevated in the BALF isolated from rats instilled 15 months earlier with 100 mg/kg fine TiO_2, as well as by 10 or 100 mg/kg of α-quartz or carbon black. *Hprt* mutations could be induced in RLE-6TN cells *in vitro* by cells from the BALF isolated from the 100 mg/kg fine TiO_2-treated rats. The authors concluded that the results supported a role for particle-elicited macrophages and neutrophils in the *in vivo* mutagenic effects of particle exposure, possibly mediated by cell-derived oxidants.

An IT study in hamsters suggested that fine TiO_2 (97% < 5 μm, including 51% < 0.5 μm) may act as a co-carcinogen [Stenbäck et al. 1976]. When BAP and fine TiO_2 (<0.5 μm particle size) were administered by IT to 48 hamsters (male and female Syrian golden hamsters, 6–7 weeks of age), 16 laryngeal, 18 tracheal, and 18 lung tumors developed, compared to only 2 laryngeal tumors found in the BAP-treated controls. In hamsters receiving an IT dose of 3 mg fine TiO_2 only in 0.2 ml saline once a week for 15 weeks, no respiratory tract tumors were found. The animals were kept until death, which occurred by 80 weeks in treated hamsters and by 120 weeks in controls.

TiO_2 was included in an IT study of 19 different dusts in female SPF Wistar rats [Pott and Roller 2005; Mohr et al. 2006]. The types of TiO_2 tested were ultrafine hydrophilic (P25; "majority anatase"; ~0.025 μm mean particle size; 52 m^2/g specific surface area), ultrafine hydrophobic (coated) (P805; 21 nm particle size; 32.5 m^2/g specific surface area), and small-fine anatase (hydrophilic) (200 nm particle size; 9.9 m^2/g specific surface area). Groups of 48 rats were administered two or three IT doses, and then maintained for 26 weeks before terminal sacrifice. The IT doses (number of doses × mass per dose) and the corresponding lung tumor response (percentages of rats with benign or malignant tumors) included ultrafine hydrophilic TiO_2 (5 doses of 3 mg, 5 doses of 6 mg, and 10 doses of 6 mg—52%, 67%, 69% tumors, respectively), ultrafine hydrophobic (coated) TiO_2 (15 doses of 0.5 mg, and 30 doses of 0.5 mg—0% and 6.7% tumors, respectively); and small-fine anatase TiO_2 (10 doses of 6 mg, and 20 doses of 6 mg—30% and 64% tumors, respectively) [Mohr et al. 2006]. The original 6-mg dose for hydrophobic coated TiO_2 was reduced to 0.5 mg because of acute mortality at the higher dose. The TiO_2 was analyzed with the tumor data for the other poorly soluble particles (1,002 rats surviving 26 weeks), and they found that the dose metric that provided the best fit to the tumor data was particle volume and particle size [Pott and Roller 2005; Roller and Pott 2006]. Borm et al. [2000] and Morfeld et al. [2006] analyzed a subset of these data (709 rats) for five different poorly soluble particles of different sizes (TiO_2 of low and high surface area, carbon black, diesel exhaust particles, and amorphous silica). Morfeld et al. [2006] fit a multivariate Cox model to these pooled data (excluding silica) and reported a threshold dose of 10 mg and a saturation dose of 20 mg for lung tumors. Although ultrafine particles were more tumorigenic than fine particles, in their multivariate model no difference was seen between particle mass, volume, or surface area after accounting for particle type and rat survival time; they suggested a high degree of agglomeration in these IT preparations may have reduced the effective particle surface area relative to that estimated from Brunauer, Emmett, and Teller (BET) analysis [Morfeld et al. 2006]. Roller [2007] considered the threshold findings to be inconsistent with the statistically significant lung tumor incidences in three dose groups that were within the 95% confidence interval of the estimated threshold in Morfeld et al. [2006]. All of the analyses of these pooled data including TiO_2 showed a greater tumor potency of the ultrafine versus fine particles, whether the dose was expressed as particle volume and size or as particle surface area [Borm et al. 2000; Pott and Roller 2005; Roller and Pott 2006; Morfeld et al. 2006]. This was considered to be due to the greater translocation of ultrafine particles to the lung interstitium [Borm et al. 2000; Pott and Roller 2005]. Although these IT studies used relatively high-mass doses, increasing dose-response relationships were observed for particles of a given size; greater tumor responses were also observed for the ultrafines compared to fine particles at a given mass dose. One study suggests a genotoxic mechanism involving DNA damage from oxidants produced by phagocytic and inflammatory cells in response to the TiO_2 particles in the lungs.

3.2.2 Acute or Subacute Inhalation

Nurkiewicz et al. [2008] investigated the role of particle size in systemic microvascular function in male SPF Sprague Dawley rats inhaling either fine TiO_2 (1 μm; 2.34 m^2/g) or P25 ultrafine TiO_2 (21 nm; 48.08 m^2/g) at airborne exposures aimed at achieving similar particle mass deposition in the lungs (ultrafine: 1.5–12 mg/m^3, 240–720 min; fine: 3–15 mg/m^3, 240–480 min). No evidence of pulmonary inflammation or lung damage (based on BALF markers) was observed at these exposures. However, 24 hours after exposure, the arteriolar vasodilation response (to Ca2+ intraluminal infusion of the spinotrapezius muscle) was found to be significantly impaired in rats exposed to ultrafine TiO_2 compared to either the control rats or the rats exposed to fine TiO_2 with the same retained mass dose in the lungs. On an equivalent mass basis, ultrafine TiO_2 was approximately an order of magnitude more potent than fine TiO_2 in causing systemic microvascular dysfunction. When converted to surface area dose, the potency of the fine TiO_2 was greater, which the authors suggested was due to overestimation of ultrafine particle surface area delivered to the lungs due to agglomeration. Either fine or ultrafine TiO_2 caused systemic microvessel dysfunction at inhalation doses that did not cause marked lung inflammation. This effect was related to the adherence of PMNs to the microvessel walls and production of ROS in the microvessels. This study indicates that cardiovascular effects may occur at particle exposure concentrations below those causing adverse pulmonary effects.

Rats (male WKY/Kyo@Rj, 246–316 g body weight) were exposed by inhalation (endotracheal intubation) to ultrafine TiO_2 (20 nm count median diameter) at an airborne mass concentration of approximately 0.1 mg/m^3 (7.2×10^6 mean number concentration) for 1 hour [Geiser et al. 2008]. BAL was performed either 1 or 24 hours after the inhalation exposure. Elemental microanalysis of the particles provided evidence that the alveolar macrophages did not efficiently phagocytize the particles; rather particle uptake was "sporadic" and "unspecific."

Mice (C57Bl/6 male, 6 weeks of age) were exposed by whole-body inhalation to TiO_2 nanoparticles (2–5 nm primary particle size; 210 m^2/g specific surface area) for either 4 hours (acute) or 4 hr/day for 10 days (subacute) [Grassian et al. 2007]. Airborne TiO_2 concentrations were 0.77 or 7.22 mg/m^3 for the acute exposure and 8.88 mg/m^3 for the subacute exposures. In the subacute study, groups of mice were necropsied at the end of the exposure period and at 1, 2, and 3 weeks postexposure. No adverse effects were observed after the 4-hour exposure. A "significant but modest" inflammatory response was observed in the mice at 0, 1, or 2 weeks after the subacute exposures, with recovery at the 3rd week postexposure (the number of alveolar macrophages in BALF was statistically significantly greater than controls in the 1- and 2-week postexposure groups).

3.2.3 Short-Term Inhalation

Short-term exposure to respirable fine TiO_2 has been shown to result in particle accumulation in the lungs of rodents inhaling relatively high particle concentrations. The pulmonary retention of these particles increased as exposure concentrations increased. In one study, after 4 weeks of exposure to 5 mg/m³, 50 mg/m³, and 250 mg/m³, the fine TiO_2 retention half-life in the lung increased (~68 days, ~110 days, and ~330 days, respectively) [Warheit et al. 1997], which indicates overloading of alveolar macrophage-mediated clearance of particles from the lungs.

In multiple studies, the most frequently noted change after 1 to 4 weeks of fine TiO_2 inhalation was the appearance of macrophages laden with particles, which were principally localized to the alveoli, bronchus-associated lymphoid tissue, and lung-associated lymph nodes [Driscoll et al. 1991; Warheit et al. 1997; Huang et al. 2001]. Particle-laden macrophages increased in number with increasing intensity of exposure and decreased in number after cessation of exposure [Warheit et al. 1997]. Alveolar macrophages from rats inhaling 250 mg/m³ fine TiO_2 for 4 weeks also appeared to be functionally impaired as demonstrated by persistently diminished chemotactic and phagocytic capacity [Warheit et al. 1997].

Inflammation in the lungs of fine rutile TiO_2-exposed rats was dependent upon exposure concentration and duration. Rats exposed to 250 mg/m³ fine TiO_2 6 hr/day, 5 days/week for 4 weeks had markedly increased numbers of granulocytes in BALF [Warheit et al. 1997]. The granulocytic response was muted after recovery, but numbers did not approach control values until 6 months after exposures ceased. Rats exposed to 50 mg/m³ fine TiO_2 6 hr/day, 5 days/wk for 4 weeks had a small but significantly increased number of granulocytes in the BALF that returned to control levels at 3 months after exposures ceased [Warheit et al. 1997]. This study showed that high concentrations of poorly soluble, low toxicity (PSLT) dust (e.g., fine TiO_2 and carbonyl iron) caused impaired pulmonary clearance and persistent inflammation in rats.

In a study of male Fischer 344 rats, inhaling 22.3 mg/m³ of ultrafine (20 nm particle size, anatase) TiO_2 6 hr/day, 5 days/wk for up to 12 weeks, the antioxidant enzyme manganese-containing superoxide dismutase (MnSOD) expression in the lungs increased dramatically and was correlated with pulmonary inflammation indicators [Janssen et al. 1994]. Fine TiO_2 (23.5 mg/m³ of 250 nm particle size, anatase) did not produce this response at the exposure conditions in this study. Follow-up observation of the rats in this study showed that the inflammatory lesions "regressed" during a 1-year period following cessation of exposure [Baggs et al. 1997]. This observation suggests that the inflammatory response from short-term exposures to TiO_2 may be reversible to some degree, if there is a cessation of exposure.

In a separate study, rats exposed to airborne concentrations of 50 mg/m³ fine TiO_2 7 hr/day, 5 days/week for 75 days had significantly elevated neutrophil numbers, LDH (a measure of cell injury) concentration, and *n*-acetylglucosaminidase (a measure of inflammation) concentration in BALF [Donaldson et al. 1990]. However, in this study the BALF of rats inhaling 10 mg/m³ or 50 mg/m³ fine TiO_2 7 hr/day, 5 days/week for 2 to 32 days had PMN numbers, macrophage numbers, and LDH concentrations that were indistinguishable from control values [Donaldson et al. 1990].

Rats exposed to airborne concentrations of 51 mg/m³ fine TiO_2 (1.0 μm mass median aerodynamic diameter [MMAD]) 6 hr/day for 5 days (whole body exposures) had no significant changes in BALF neutrophil number, macrophage number, lymphocyte number, LDH concentration, *n*-acetylglucosaminidase concentration, or measures of macrophage activation 1 to 9 weeks after exposure [Driscoll et al. 1991]. The TiO_2 lung burden at the end of exposure was 1.8 mg/lung, and the retention was 39% 28 days after the end of exposure. Similarly, rats exposed to 0.1, 1, or 10 mg/m³ 6 h/day, 5 days/week for 4 weeks had no evidence of lung injury as assessed by BAL 1 week to 6 months after exposure or by histopathology 6 months after exposure [Henderson et al. 1995].

Pulmonary responses to six different formulations of fine rutile TiO_2 (including uncoated or alumina- or amorphous silica-coated; particle size 290–440 nm; specific surface area 6–28 m²/g) were investigated in male, 8-week old Sprague Dawley rats, with both IT (see Section 3.2.1.1) and inhalation exposures. Rats were exposed to very high airborne concentrations (1130–1310 mg/m³) of the different formulations of fine TiO_2 for 30 days (6 hr/day, 5 days/week). The pulmonary inflammation response (assessed by histopathology) remained significantly elevated 1 month after exposure to the TiO_2 coated with alumina (7%) and amorphous silica (8%). The coated TiO_2 formulations produced higher inflammation than the uncoated TiO_2 in the inhalation study (as in the IT study) [Warheit et al. 2005].

3.2.4 Subchronic Inhalation

In a study of two fine-sized, PSLT particles—TiO_2 and barium sulfate ($BaSO_4$)—no-observed-adverse-effect levels (NOAELs) were estimated based on the relationship between the particle surface area dose, overloading of lung clearance, and neutrophilic inflammation in rats [Tran et al. 1999]. These two PSLT particles had similar densities (4.25 and 4.50 g/cm³, respectively) but different particle sizes (2.1 and 4.3 μm MMAD); since both of these factors influence particle deposition in the lungs, the exposure concentrations were adjusted to provide similar particle mass deposition in the lungs. Male Wistar rats (age 12 weeks, specific-pathogen free) were exposed by whole body inhalation (7 hr/day, 5 days/wk) to either 25 mg/m³ for 7.5 months (209 calendar days) or to 50 mg/m³ for 4 months (118 calendar days). The findings showed that the retardation of alveolar macrophage-mediated clearance, particle transfer to the lung-associated lymph nodes, and influx of PMNs were related to the lung burden as particle surface area dose. A mean airborne concentration of 3 mg/m³ fine-sized TiO_2 was estimated as the NOAEL, which was defined as a 95% probability that the lung responses would be below those predicted using the "no overload level" for the average animal.

The relationship between subchronic inhalation of TiO_2 and neutrophilic inflammation was also investigated in a study of Montserrat volcanic ash [Cullen et al. 2002]. Male Wistar rats (225 g, specific-pathogen free) were exposed by nose-only inhalation (6 hr/day, 5 days/wk) to 140 mg/m³ of TiO_2 (1.2 μm MMAD) for up to 2 months. The concentration of ash (5.3 µm MMAD) was 253 mg/m³, which was predicted to provide the same retained lung burden as the TiO_2. After the 8-week exposure, the histopathological examination showed relatively minor pathology changes in the lungs of rats treated with TiO_2, although the pathological changes in the lungs of the ash-exposed rats were generally "more marked." The pulmonary inflammation response was also greater in the ash-exposed rats, based on the percentage of PMNs in the BALF, which ranged from 19%–53% PMNs for the ash and from 0.2%–16% for TiO_2 at 14–56 days of exposure.

Fine TiO_2 (pigmentary; 1.4 µm MMAD) was studied in female rats (CDF[F344]/Crl-BR), mice (B3C3F1/CrlBR), and hamsters [Lak:LVG(SYR)BR], 6 wk old, after a 13-week inhalation exposure to 10, 50, or 250 mg/m³, followed by up to 52 weeks without TiO_2 exposure [Bermudez et al. 2002; Everitt et al. 2000]. Retained particle burdens in the lungs and lung-associated lymph nodes were measured at the end of exposure and at 4, 13, 26, and 52 weeks postexposure. The lung particle retention patterns indicated that after the 13 weeks of exposure to 50 or 250 mg/m³, clearance overload had occurred in rats and mice but not in hamsters. In mice and rats, the numbers of macrophages and the percentage of neutrophils were significantly increased in BALF after 13 weeks of exposure to 50 or 250 mg/m³ fine TiO_2 and remained elevated through the 52-week postexposure. These BALF cell responses were significantly elevated in hamsters after exposure to 250 mg/m³ but were no longer significantly elevated by 26 weeks postexposure.

Alveolar cell proliferation was significantly elevated only at 52-week postexposure in the rats [Bermudez et al. 2002]. Histopathology showed alveolar hypertrophy and hyperplasia of type II epithelial cells in rats after the 13-week exposure to 50 or 250 mg/m³ fine TiO_2. In mice, alveolar type II cell hypertrophy was observed (dose not given), while in hamsters minimal type II epithelial cell hypertropy and hyperplasia were observed at 50 or 250 mg/m³. Foci of alveolar epithelial cell hypertrophy and hyperplasia were seen to be often associated with aggregates of particle-laden alveolar macrophages in all three species. In rats, but not mice and hamsters, these foci of alveolar epithelial hypertrophy became increasingly more prominent with time, even after cessation of exposure, and at the high dose, rats progressed to bronchiolization of alveoli (metaplasia) and to fibrotic changes with focal interstitialization of TiO_2 particles. Alveolar lipoproteinosis and cholesterol clefts were observed in rats 52 weeks after the 13-week exposure to 250 mg/m³. Although "high particle burdens" were associated with "proliferative epithelial changes associated with particle-induced inflammation" in all three species, only rats developed metaplastic and fibrotic lesions [Everitt et al. 2000; Bermudez et al. 2002].

P25 ultrafine TiO_2 (21 nm primary particle size; 1.37 MMAD) was studied in female rats, mice, and hamsters after a 13-week inhalation exposure to 0.5, 2, or 10 mg/m³, followed by a recovery period (without TiO_2 exposure) for up to 52 weeks [Bermudez et al. 2004] (same rodent strains as Bermudez et al. 2002). Pulmonary responses and retained particle burdens in the lungs and lung-associated lymph nodes were measured at the end of exposure and at 4, 13, 26, and 52 weeks postexposure. Retardation of pulmonary clearance following exposure to 10 mg/m³ was observed in rats and mice but not in hamsters. Pulmonary inflammation was also observed at the 10 mg/m³ dose in rats and mice but not in hamsters. The total number of cells in BALF was significantly elevated in rats and mice after 13 weeks of exposure to 10 mg/m³, and the percentages of neutrophils and macrophages remained statistically significantly elevated through 52 weeks postexposure in both species. The BALF cell responses were not statistically significant following exposure to 0.5 or 2 mg/m³ in mice or rats (except for significantly elevated neutrophils in rats immediately after the 13-wk exposure to 2 mg/m³) (Tables S1–S3 in Bermudez et al. 2004). BALF cell responses were not significantly elevated in hamsters at any exposure or observation time.

The alveolar cell replication index was statistically significantly increased at 0, 4, and 13 weeks after the 13-wk exposure to 10 mg/m³ ultrafine TiO_2 in rats, while in mice it was significantly increased at 13 and 26 weeks after exposure cessation [Bermudez et al. 2004]. In rats inhaling 10 mg/m³, the histopathologic responses included epithelial and fibroproliferative changes, interstitial particle accumulation, and alveolar septal fibrosis. Although most of the epithelial proliferative lesions had regressed postexposure, some remaining lesions were believed to have progressed. At 52 weeks postexposure, minimal to mild metaplastic changes and minimal to mild particle-induced alveolar septal fibroplasia were seen in rats. In mice inhaling 10 mg/m³, lesions were described as aggregations of heavily particle laden macrophages in the centriacinar sites. During postexposure, these cell aggregates were observed to move to the interstitial areas over time. No epithelial, metaplastic, and fibroproliferative changes were observed by histopathology in the mice or hamsters (although the mice had significantly elevated alveolar cell replication at 13 and 26 weeks postexposure in the 10 mg/m³ dose group, this response apparently did not result in histopathologically visible changes). The absence of adverse pulmonary responses in hamsters was considered to reflect their rapid clearance of particles from the lungs.

3.2.5 Chronic Inhalation

TiO_2 has been investigated in three chronic inhalation studies in rats, including fine TiO_2 in Lee et al. [1985] and Muhle et al. [1991] and ultrafine TiO_2 in Heinrich et al. [1995]. These studies were also reported in other publications, including Lee et al. [1986a], Muhle et al. [1989, 1994], and Bellmann et al. [1991]. In another 2-year rat inhalation study, an increase in lung tumors was found in rats exposed to $TiCl_4$ [Lee et al. 1986b]. $TiCl_4$ is an intermediate in the production of pigment TiO_2, including by hydrolysis of $TiCl_4$ to produce TiO_2 and HCl. However, $TiCl_4$ is a highly volatile compound with different properties than TiO_2, and thus it is not addressed further in this document.

In Lee et al. [1985], groups of 100 male and 100 female rats (CD, Sprague-Dawley derived; strain not specified) were exposed by whole body inhalation to fine rutile TiO_2 (1.5–1.7 μm MMAD) for 6 hr/day, 5 days/week, for up to two years, to 10, 50, or 250 mg/m³ (84% respirable; < 13 μm MMAD). A fourth group (control) was exposed to air. In each group, twenty rats were killed at 3, 6, or 12 months; 80 rats were exposed for 2 years, and all surviving rats were killed at the end of exposure. No increase in lung tumors was observed at 10 or 50 mg/m³. At 250 mg/m³, bronchioalveolar adenomas were observed in 12 out of 77 male rats and 13 out of 74 female rats. In addition, squamous cell carcinomas were reported in 1 male and 13 females at 250 mg/m³. The squamous cell carcinomas were noted as being dermoid, cyst-like squamous cell carcinomas [Lee et al. 1985], later reclassified as proliferative keratin cysts [Carlton 1994], and later still as a continuum ranging from pulmonary keratinizing cysts through pulmonary keratinizing eptheliomas to frank pulmonary squamous carcinomas [Boorman et al. 1996]. A recent reanalysis of the 16 tumors originally classified as cystic keratinizing squamous cell carcinomas in Lee et al. [1985] had a similar interpretation: two were reclassified as squamous metaplasia, one as a poorly keratinizing squamous cell carcinoma, and 13 as nonneoplastic pulmonary keratin cysts [Warheit and Frame 2006].

In both the Muhle et al. [1991] and Heinrich et al. [1995] studies, TiO_2 was used as a negative control in 2-year chronic inhalation studies of toner and diesel exhaust, respectively. In Muhle et al. [1991], the airborne concentration of TiO_2 (rutile) was 5 mg/m³ (78% respirable, according to 1984 ACGIH criterion). Male and female Fischer 344 rats were exposed for up to 24 months by whole body inhalation and sacrificed beginning at 25.5 months. No increase in lung tumors was observed in TiO_2-exposed animals; the lung tumor incidence was 2/100 in TiO_2-exposed animals versus 3/100 in nonexposed controls.

In the Heinrich et al. [1995] study, 100 female Wistar rats were exposed to ultrafine TiO_2 (80% anatase, 20% rutile; 15–40 nm primary particle size; 0.8 µm MMAD; 48 (± 2.0) m²/g specific surface area) at an average of approximately 10 mg/m³, 18 h/day 5d/wk, for up to 24 months (actual concentrations were 7.2 mg/m³ for 4 months, followed by 14.8 mg/m³ for 4 months, and 9.4 mg/m³ for 16 months). Following the 2-year exposure, the rats were held without TiO_2 exposure for 6 months [Heinrich et al. 1995]. At 6 months of exposure, 99/100 of the rats had developed bronchioloalveolar hyperplasia, and by 2 years all rats had developed slight to moderate interstitial fibrosis [Heinrich et al. 1995]. After 24 months of exposure, four of the nine rats examined had developed tumors (including a total of two squamous cell carcinomas, one adenocarcinoma, and two benign squamous cell tumors). At 30 months (6 months after the end of exposure), a statistically significant increase in adenocarcinomas was observed (13 adenocarcinomas, in addition to 3 squamous cell carcinomas and 4 adenomas, in 100 rats). In addition, 20 rats had benign keratinizing cystic squamous-cell tumors. Only 1 adenocarcinoma, and no other lung tumors, was observed in 217 nonexposed control rats.

NMRI mice were also exposed to ultrafine TiO_2 in Heinrich et al. [1995]. The lifespan of NMRI mice was significantly decreased by inhaling approximately 10 mg/m³ ultrafine TiO_2, 18 hr/day for 13.5 months [Heinrich et al. 1995]. This exposure did not produce an elevated tumor response in the NMRI mice, but the 30% lung tumor prevalence in controls may have decreased the sensitivity for detecting carcinogenic effects in this assay. In a study of several types of particles, 100 female Wistar (Crl:[WI] Br) rats were exposed to 10.4 mg/m³ TiO_2, 18 hr/day 5 days/wk, for 24 months (followed by 6 months in clean air). No information was provided on the particle size or crystal structure of TiO_2 used in this study. Cystic keratizing epitheliomas were observed in 16% of the rats. In addition, 3.0% cystic keratinizing squamous-cell carcinomas and 1% nonkeratinizing squamous-cell carcinoma were observed [Rittinghausen et al. 1997].

The primary data used in the dose-response model in the TiO_2 risk assessment (Chapter 4) include the ultrafine TiO_2 data of Heinrich et al. [1995] and the fine TiO_2 data of Lee et al. [1985]. Differences in postexposure follow-up (24 and 30 months, respectively, in Lee et al. [1985] and Heinrich et al. [1995]) may have increased the likelihood of detecting lung tumors in the ultrafine TiO$_2$-exposed rats. However, the differences in the hours of exposure per day (6 and 18 hrs, respectively, in Lee et al. [1985] and Heinrich et al. [1995])) were accounted for in the risk assessment models since the retained particle lung burden (at the end of 2-year inhalation exposure) was the dose metric used in those models.

In summary, the chronic inhalation studies in rodents show dose-related pulmonary responses to fine or ultrafine TiO$_2$. At sufficiently high particle mass or surface area dose, the responses in rats include reduced lung clearance and increased particle retention ("overload"), pulmonary inflammation, oxidative stress, tissue damage, fibrosis, and lung cancer. Studies in mice showed impaired lung clearance and inflammation, but not fibrosis or lung cancer. Studies in hamsters found little adverse effect of TiO$_2$ on either lung clearance or response.

3.3 In Vivo Studies: Other Routes of Exposure

3.3.1 Acute Oral Administration

The acute toxicity of nanometer (25 and 80 nm) and submicron (155 nm) TiO$_2$ was investigated after a large single dose of TiO$_2$ (5 g/kg body weight) in male and female CD-I mice [Wang et al. 2007b]. The TiO$_2$ was retained in liver, spleen, kidney, and lung tissues, indicating uptake by the gastrointestinal tract and systemic transport to the other tissues. No acute toxicity was observed. However, statistically significant changes were seen in several serum biochemical parameters, including LDH and alpha-hydroxybutyrate (suggesting cardiovascular damage), which were higher in mice treated with either 25 and 80 nm nanoscale TiO$_2$ compared to fine TiO$_2$. Pathological evidence of hepatic injury and kidney damage was also observed. In female mice treated with the 80 nm nanoscale and the fine particles, the liver injury included hydropic degeneration around the central vein and spotty necrosis of hepatocytes; renal damage included protein-filled liquid in the renal tubule and swelling of the renal glomerulus. The tissue injury markers did not always relate to particle size; damage was often but not always greater in the mice treated with the nanoscale particles compared to the fine particles, and the 80-nm TiO$_2$ was more damaging than the 25-nm TiO$_2$ by some indicators. Given the single large dose in this study, it was not possible to evaluate the dose-response relationship or to determine whether the effects were specific to the high dose.

3.3.2 Chronic Oral Administration

The National Cancer Institute conducted a bioassay of TiO$_2$ for possible carcinogenicity by the oral route. TiO$_2$ was administered in feed to Fischer 344 rats and B6C3F$_1$ mice. Groups of 50 rats and 50 mice of each sex were fed either 25,000 or 50,000 parts per million TiO$_2$ (2.5% or 5%) for 103 weeks and then observed for an additional week. In the female rats, C-cell adenomas or carcinomas of the thyroid occurred at an incidence of 1 out of 48 in the control group, 0 out of 47 in the low-dose group, and 6 out of 44 in the high-dose group. It should also be noted that a similar incidence of C-cell adenomas or carcinomas of the thyroid, as observed in the high-dose group of the TiO$_2$ feeding study, has been seen in control female Fischer 344 rats used in other studies. No significant excess tumors occurred in male or female mice or in male rats. It was concluded that, under the conditions of this bioassay, TiO$_2$ is not carcinogenic by the oral route for Fischer 344 rats or B6C3F$_1$ mice [NCI 1979].

In a study of male and female Fischer 344 rats fed diets containing up to 5% TiO_2-coated mica for up to 130 weeks, no treatment-related toxicologic or carcinogenic effects were reported [Bernard et al. 1990].

3.3.3 Intraperitoneal Injection

Female Wistar rats received intraperitoneal injections of P25 ultrafine anatase TiO_2 (2 ml 0.9% NaCl solution) of either (1) a total dose of 90 mg per animal (once per week for five weeks); (2) a single injection of 5 mg; or (3) a series of injections of 2, 4, and 4 mg at weekly intervals [Pott et al. 1987]. Controls received a single injection of saline alone. The average lifespan of rats in the three treatment groups were 120, 102, and 130, respectively, and 120 weeks for controls. In the first treatment group, 6 out of 113 rats developed sarcomas, mesotheliomas, or abdominal cavity carcinomas, compared to 2 carcinomas in controls. No intra-abdominal tumors were found in the other two treatment groups. This study may not be relevant to inhalation exposure to TiO_2 since studies have not shown that either fine or ultrafine TiO_2 would be likely to reach the peritoneum after depositing in the lungs.

3.4 Particle-Associated Lung Disease Mechanisms

3.4.1 Role of Pulmonary Inflammation

Chronic pulmonary inflammation is characterized by persistent elevation of the number of PMNs (measured in BALF) or by an increased number of inflammatory cells in interstitial lung tissue (observed by histopathology). Pulmonary inflammation is a defense mechanism against foreign material in the lungs. PMNs are recruited from the pulmonary vasculature in response to chemotactic stimuli (cytokines) generated by lung cells including alveolar macrophages, which patrol the lungs as part of the normal lung defense to phagocytose and clear foreign material [Lehnert 1993; Snipes 1996]. Additional alveolar macrophages are recruited from blood monocytes into the lung alveoli in response to particle deposition. Typically, the PMN response is short-lived but may become chronic in response to persistent stimuli, e.g., prolonged particle exposure [Lehnert 1993].

Particle-induced pulmonary inflammation, oxidative stress, lung tissue damage, and epithelial cell proliferation are considered to be the key steps leading to lung tumor development in the rat acting through a secondary genotoxic mechanism [Knaapen et al. 2004; Baan 2007]. Oxidative stress is considered the underlying mechanism of the proliferative and genotoxic responses to poorly soluble particles including TiO_2 and other PSLT [Donaldson et al. 1996; Shi et al. 1998; Vallyathan et al. 1998; Knaapen et al. 2002; Donaldson and Stone 2003]. Reactive oxygen or nitrogen species (ROS/RNS) are released by inflammatory cells (macrophages and PMNs) and/or by particle reactive surfaces. Oxidative stress results from an imbalance of the damaging oxidants and the protective antioxidants. Oxidants can damage the lung epithelial tissue and may also induce genetic damage in proliferating epithelial cells, increasing the probability of neoplastic transformation [Driscoll et al. 1997]. The mechanisms linking pulmonary inflammation with lung cancer may involve (1) induction of *hprt* mutations in DNA of lung epithelial cells by inflammatory cell-derived oxidants and increased cell proliferation [Driscoll et al. 1996], and (2) inhibition of the nucleotide excision repair of DNA (with adducts from other exposures such as polycyclic aromatic hydrocarbons) in the lung epithelial cells [Güngör et al. 2007]. Both of these mechanisms involve induction of cell-derived inflammatory mediators, without requiring particle surface-generated oxidants.

A secondary genotoxic mechanism would, in theory, involve a threshold dose that triggers inflammation and overwhelms the body's antioxidant and DNA repair mechanisms [Greim and Ziegler-Skylakakis 2007]. Antioxidant defense responses vary across species, and interindividual variability is generally greater in humans than in laboratory animals [Slade et al. 1985, 1993]. Limited quantitative data in humans make the identification of a population-based threshold difficult in practice [Bailer et al. 1997], and the quantitative aspects of the inflammatory response (level and duration) that are sufficient to cause a high probability of lung tumor development are not known. While there is a clear association between inflammation and genotoxicity, the specific linkages between key cellular processes such as cell cycle arrest, DNA repair, proliferation, and apoptosis are not well understood [Schins and Knaapen 2007].

Thus, chronic pulmonary inflammation appears to be required in the development of lung tumors in rats following chronic inhalation exposure to TiO_2; i.e., acting through a secondary genotoxic mechanism involving oxidative DNA damage. An implication of this mechanism is that maintaining exposures below those causing inflammation would also prevent tumor development, although the distribution of inflammatory responses in human populations is not known and a direct genotoxic mechanism cannot be ruled out for discrete nanoscale TiO_2 (see Section 3.5.2.1).

3.4.2 Dose Metric and Surface Properties

High mass or volume dose of PSLT fine particles in the lungs has been associated with overloading, while ultrafine particles impair lung clearance at lower mass or volume doses [Bellmann et al. 1991; Morrow et al. 1991; Oberdörster et al. 1994b; Heinrich et al. 1995]. The increased lung retention and inflammatory response of ultrafine PSLT particles compared to fine PSLT particles correlates better to the particle surface area dose [Tran et al. 2000]. Some evidence suggests that reduced lung clearance of ultrafine particles may involve mechanisms other than high-dose overloading, such as altered alveolar macrophage function (phagocytosis or chemotaxis) [Renwick et al. 2001, 2004] and greater ability to enter the lung interstitium [Ferin et al. 1992; Adamson and Bowden 1981; Oberdörster et al. 1994b; Adamson and Hedgecock 1995].

The quantitative relationships between the particle dose (expressed as mass or surface area) and the pulmonary responses of inflammation or lung tumors can be determined from the results of subchronic or chronic inhalation studies in rats. When the rat lung dose is expressed as particle mass, several different dose-response relationships are observed for pulmonary inflammation following subchronic inhalation of various types of poorly soluble particles (Figure 3–1). However, when dose is converted to particle surface area, the different types (TiO_2 and $BaSO_4$) and sizes (ultrafine and fine TiO_2) of PSLT particles can be described by the same dose-response curve (Figure 3–2), while crystalline silica, SiO_2 (a high-toxicity particle) demonstrates a more inflammogenic response when compared to PSLT particles of a given mass or surface area dose.

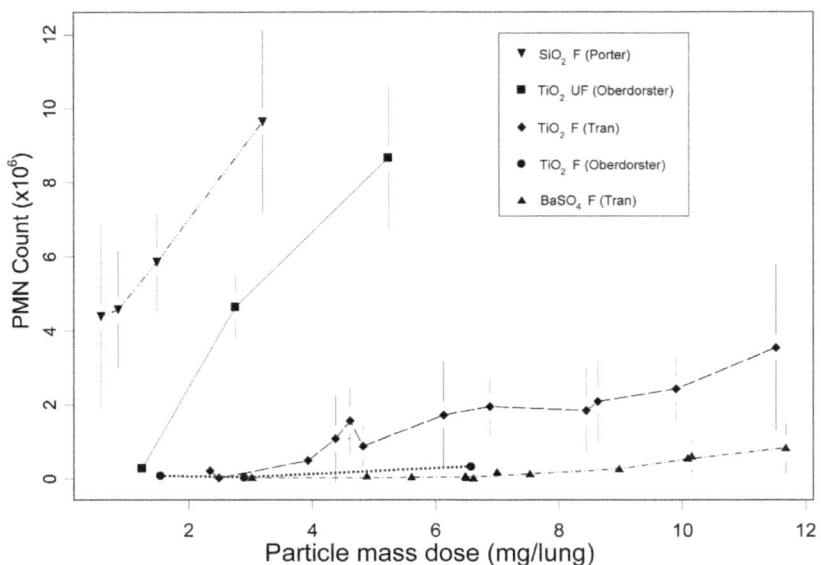

Figure 3–1. Pulmonary inflammation (PMN count) of high toxicity dust (crystalline silica) particles compared to low toxicity dust (TiO_2 and $BaSO_4$) of both fine and ultrafine size, based on particle mass dose in rat lungs; Particle size: F (fine), UF (ultrafine)

Data source: Porter et al. [2001], Oberdörster et al. [1994a], Tran et al. [1999]

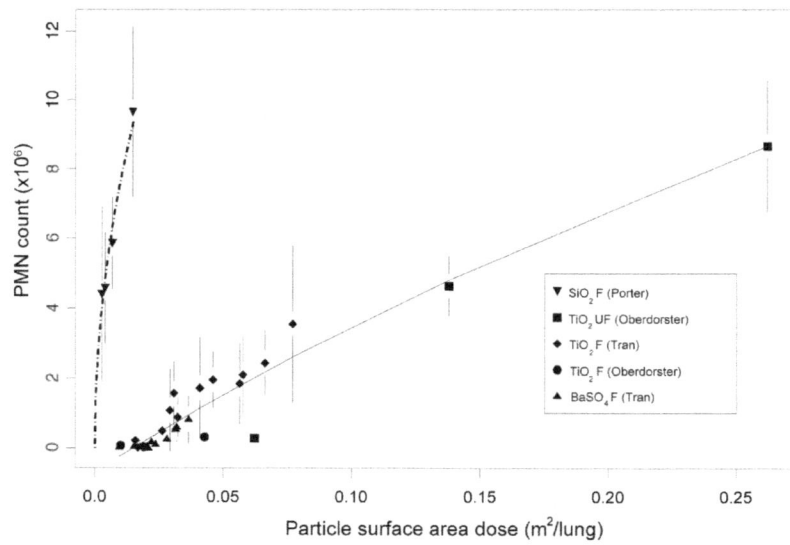

Figure 3–2. Pulmonary inflammation (PMN count) of high toxicity dust (crystalline silica) particles compared to low toxicity dust (TiO_2 and $BaSO_4$) of both fine and ultrafine size, based on particle surface area dose in rat lungs; Particle size: F (fine), UF (ultrafine)

Data source: Porter et al. [2001], Oberdörster et al. [1994a], Tran et al. [1999]

Similarly, the rat lung tumor response to various types and sizes of respirable particles (including fine and ultrafine TiO$_2$, toner, coal dust, diesel exhaust particulate, carbon black, and talc) has been associated with the total particle surface area dose in the lungs (Figure 3–3). This relationship, shown by Oberdörster and Yu [1990], was extended by Driscoll [1996] to include results from subsequent chronic inhalation studies in rats exposed to PSLT particles and by Miller [1999] who refit these data using a logistic regression model. The lung tumor response in these analyses is based on all lung tumors, since some of the studies did not distinguish the keratinizing squamous cell cysts from the squamous cell carcinomas [Martin et al. 1977; Lee et al. 1985; Mauderly et al. 1987]. Keratinizing squamous cell cysts have been observed primarily in the lungs of female rats exposed to PSLT, including TiO$_2$ (discussed further in Sections 3.2.5, 3.5.2.4, and 3.6).

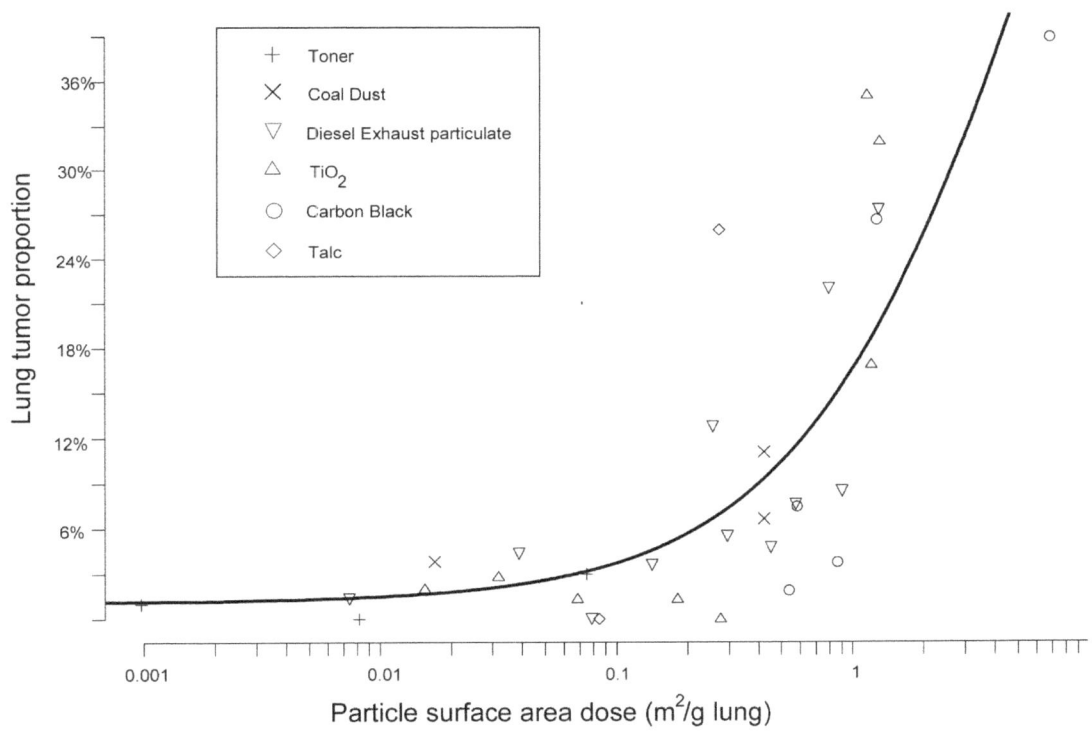

Figure 3–3. Relationship between particle surface area dose in the lungs of rats after chronic inhalation to various types of poorly soluble, low toxicity (PSLT) particles and tumor proportion (all tumors including keratinizing squamous cell cysts)

Data source: toner [Muhle et al. 1991], coal dust [Martin et al. 1977], diesel exhaust particulate [Mauderly et al. 1987; Lewis et al. 1989; Nikula et al. 1995; Heinrich et al. 1995], TiO$_2$ [Muhle et al. 1991; Heinrich et al. 1995; Lee et al. 1985, 1986a], carbon black [Nikula et al. 1995; Heinrich et al. 1995], talc [NTP 1993]

Figures 3–4 and 3–5 show the particle mass and surface area dose-response relationships for fine and ultrafine TiO$_2$ for chronic inhalation exposure and lung tumor response in rats. In these figures, the lung tumor response data are shown separately for male and female rats at 24 months in Lee et al. [1985] and for female rats at 24 or 30 months, including either all tumors or tumors without keratinizing cystic tumors [Heinrich et al. 1995] since this study distinguished these tumor types. The data are plotted per gram of lung to adjust for differences in the lung mass in the two strains of rats (Sprague-Dawley and Wistar). Figure 3–4 shows that when TiO$_2$ is expressed as mass dose, the lung tumor response to ultrafine TiO$_2$ is much greater at a given dose than that for fine TiO$_2$; yet when TiO$_2$ is expressed as particle surface area dose, both fine and ultrafine TiO$_2$ data fit the same dose-response curve (Figure 3–5). These findings indicate that, like other PSLT particles, TiO$_2$ is a rat lung carcinogen, and for a given surface area dose, the equivalent mass dose associated with elevated tumor response would be much higher for fine particles than for ultrafine particles.

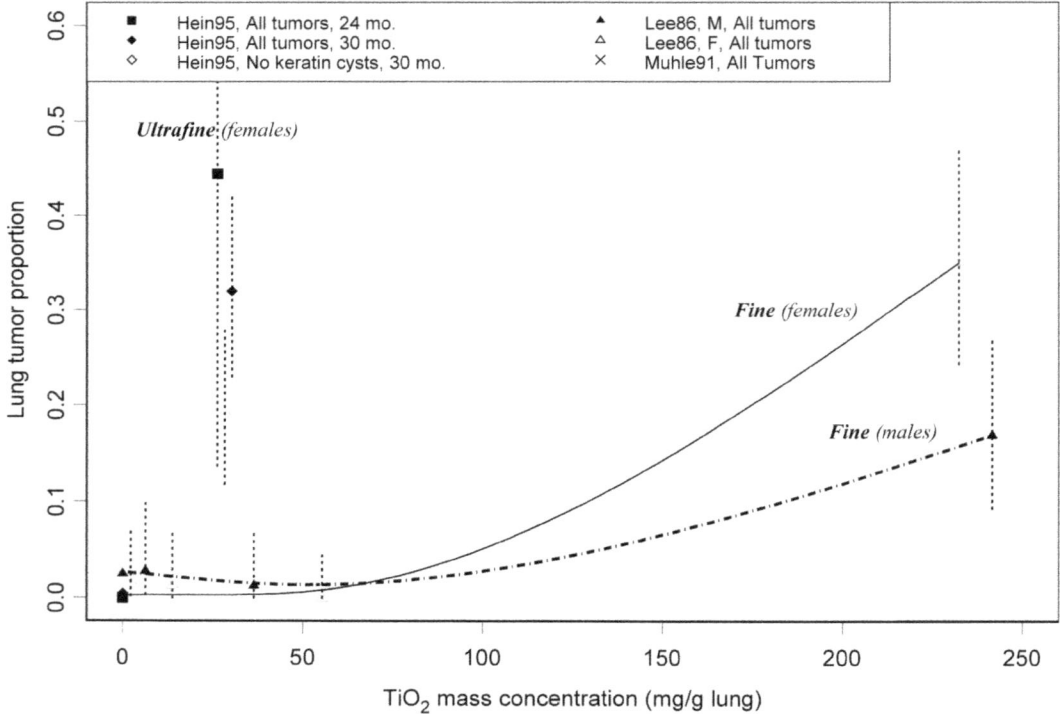

Figure 3–4. TiO$_2$ mass dose in the lungs of rats exposed by inhalation for 2 years and tumor proportion (either all tumors or tumors excluding keratinizing squamous cell cysts)

Note: Spline model fits to Lee data. (Heinrich dose data are jittered, i.e., staggered)

Data source: Heinrich et al. [1995], Lee et al. [1985, 1986a], and Muhle et al. [1991]

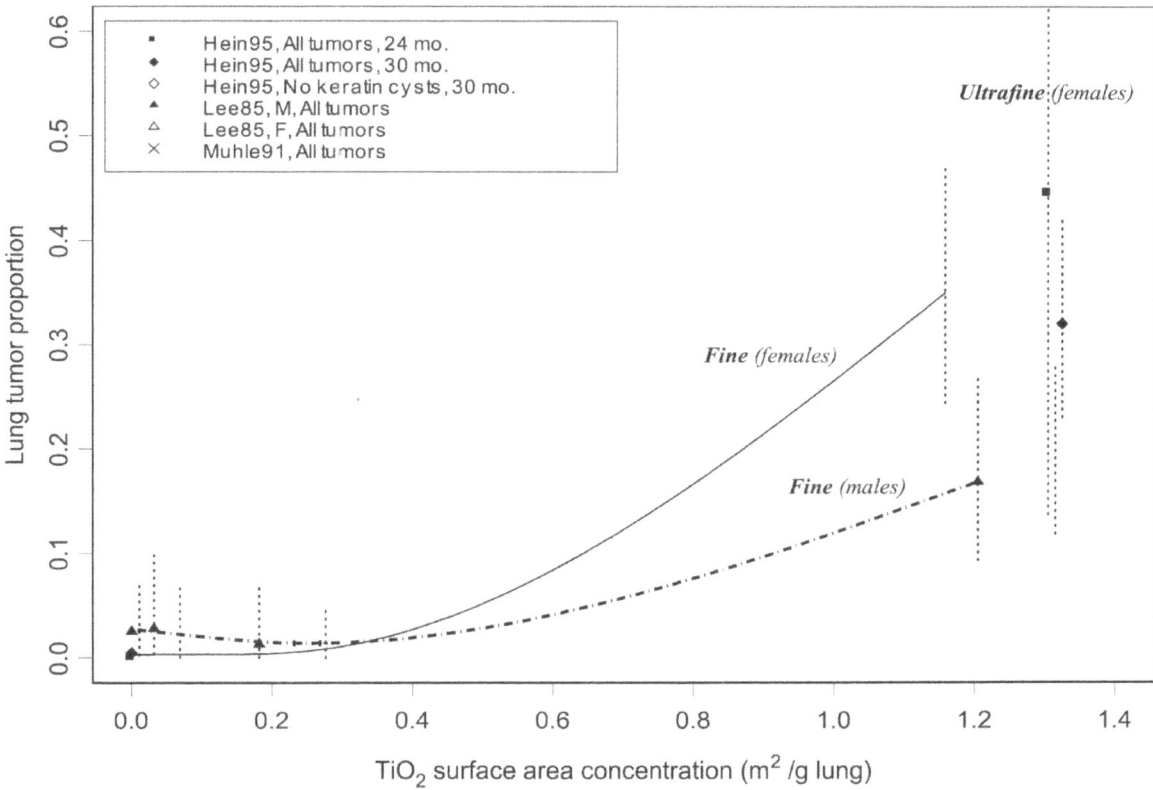

Figure 3–5. TiO$_2$ surface area dose in the lungs of rats exposed by inhalation for 2 years and tumor proportion (either all tumors or tumors excluding keratinizing squamous cell cysts)

Note: Spline model fits to Lee data. (Heinrich dose data are jittered, i.e., staggered)

Data source: Heinrich et al. [1995], Lee et al. [1985, 1986a], and Muhle et al. [1991]

The difference in TiO$_2$ crystal structure in these subchronic and chronic studies did not influence the dose-response relationships for pulmonary inflammation and lung tumors [Bermudez et al. 2002, 2004; Lee et al. 1985; Heinrich et al. 1995]. That is, the particle surface area dose and response relationships were consistent for the ultrafine (80% anatase, 20% rutile) and fine (99% rutile) TiO$_2$ despite the differences in crystal structure. In contrast, differences in ROS generation and toxicity have been observed for different TiO$_2$ crystal structures in cell-free, *in vitro*, and short-term *in vivo* studies. Cell-free assays have reported crystal structure (anatase, rutile, or mixtures) influences particle surface ROS generation [Kawahara et al. 2003; Kakinoki et al. 2004; Behnajady et al. 2008; Jiang et al. 2008]. In a cell-free study designed to investigate the role of surface area and crystal structure on particle ROS generation, Jiang et al. [2008] observed that size, surface area, and crystal structure all contribute to ROS generation. The ROS generation was associated with the number of defective sites per surface area, which was constant for many of the particle sizes but varied for some of the smaller particle sizes (10–30 nm) due to differences in particle generation methods. In an *in vitro* cell assay, cytotoxicity was associated with greater ROS generation from photoactivated TiO$_2$ [Sayes et al. 2006]. In a pulmonary toxicity study in rats, the surface activity of the TiO$_2$ particle (related to crystal structure, passivation, and acidity) was associated with the inflammation and cell proliferation responses at early time points, but not at the later time points [Warheit et al. 2007].

Although these studies (cited in preceding paragraph) indicate that the particle surface properties pertaining to the crystal structure of TiO$_2$, including photoactivation, can influence the ROS generation, cytotoxicity, and acute lung responses, these studies also show that crystal structure does not influence the pulmonary inflammation or tumor responses following subchronic or chronic exposures. The reason for these differences in the acute and longer-term responses with respect to TiO$_2$ crystal structure are not known but could relate to immediate effects of exposure to photoactivated TiO$_2$ and to quenching of ROS on the TiO$_2$ surfaces by lung surfactant. Unlike crystalline silica, which elicits a pronounced pulmonary inflammation response, the inflammatory response to TiO$_2$ of either the rutile or anatase/rutile crystal form was much less pronounced after subchronic exposure in rats [Bermudez et al. 2002, 2004] (Figure 3-2).

Reactive species (ROS/RNS) are also produced by alveolar macrophages and inflammatory cells during lung clearance and immunological responses to inhaled particles, and the oxidative stress that can occur when antioxidant defenses are overwhelmed is considered an underlying mechanism of the proliferative and genotoxic responses to inhaled particles [Donaldson et al. 1996; Shi et al. 1998; Vallyathan et al. 1998; Knaapen et al. 2002, 2004; Donaldson and Stone 2003; Baan 2007]. PSLT particles, including TiO$_2$, have relatively low surface reactivity compared to the more inherently toxic particles with higher surface activity such as crystalline silica [Duffin et al. 2007]. These findings are based on the studies in the scientific literature and may not apply to other formulations, surface coatings, or treatments of TiO$_2$ for which data were not available.

3.5 Particle-Associated Lung Responses

3.5.1 Rodent Lung Responses to Fine and Ultrafine TiO$_2$

Like other PSLT, fine and ultrafine TiO$_2$ can elicit persistent pulmonary inflammation at sufficiently high dose and/or duration of exposure. This occurs at doses that impair the normal clearance of particles from the alveolar (gas exchange) region of the lungs, i.e., "overloading" of alveolar macrophage-mediated particle clearance from the lungs [ILSI 2000; Mauderly 1996; Morrow 1988; Le Bouffant 1971]. As the lung burden increases, alveolar macrophages become activated and release ROS/RNS and cellular factors that stimulate pathological events [Driscoll et al. 1990; Lapp and Castranova 1993; Castranova 1998; Driscoll 2000]. Lung overload is characterized in rats, and to some extent in mice and hamsters, as increased accumulation of particle-laden macrophages, increased lung weight, infiltration of neutrophils, increased epithelial permeability, increased transfer of particles to lymph nodes, persistent inflammation, lipoproteinosis, fibrosis, alveolar epithelial cell hyperplasia, and (eventually in rats) metaplasia and nonneoplastic and neoplastic tumors [ILSI 2000; Mauderly 1996; Morrow et al. 1991; Muhle et al. 1991].

Lung inflammation and histopathological responses to TiO$_2$ and other PSLT particles were more severe in the rat than in the mouse or hamster strains studied [Everitt et al. 2000; Bermudez et al. 2002, 2004; Elder et al. 2005; Heinrich et al. 1995]. Qualitatively similar early lung responses have been observed in rodents, especially rats and mice, although differences in disease progression and severity were also seen [Everitt et al. 2000; Bermudez et al. 2002, 2004; Elder et al. 2005]. Hamsters continued to effectively clear particles from the lungs, while the mice and rats developed overloading of lung clearance and retained higher lung particle burdens postexposure [Bermudez et al. 2002, 2004; Elder et al. 2005]. Mice and rats developed persistent pulmonary inflammation through 52 weeks after a 13-week exposure to 50 or 250 mg/m³ fine TiO$_2$ [Everitt et al. 2000; Bermudez et al. 2002] or to 10 mg/m³ ultrafine TiO$_2$ [Bermudez et al. 2004]. The rat lung response to another PSLT (carbon black) was more inflammatory (e.g., greater and more sustained generation of ROS in lung cells and higher levels of some inflammatory cytokines in BALF) compared to mice or hamsters [Carter et al. 2006]. All three rodent species developed proliferative epithelial changes at the higher doses; however, only the rat developed metaplasia and fibrosis [Everitt et al. 2000; Bermudez et al. 2002, 2004]. Both rats and mice inhaling P25 ultrafine TiO$_2$ had elevated alveolar cell proliferation [Bermudez et al. 2004], but only rats had a significantly elevated lung tumor response [Heinrich et al. 1995].

Different strains of mice were used in the subchronic and chronic studies (B3C3F1/CrlBR in Bermudez et al. [2004] and NMRI in Heinrich et al. [1995]). The NMRI mice had a high background tumor response (in the unexposed control mice), which might have limited the ability to detect any particle-related increase in that study [Heinrich et al. 1995]. The B3C3F1/CrlBR mouse strain, which showed a proliferative alveolar cell response to ultrafine TiO$_2$ [Bermudez et al. 2004], has not been evaluated for tumor response to TiO$_2$. Thus, the mouse data are limited as to their utility in evaluating a TiO$_2$ lung tumor response. In addition, the rapid particle lung clearance in hamsters and their relative insensitivity to inhaled particles also limits the usefulness of that species in evaluating lung responses to TiO$_2$. In the rat studies, three different strains were used (Sprague-Dawley [Lee et al. 1985], Wistar [Heinrich et al. 1995], and Fischer-344 [Muhle et al. 1991]), although no heterogeneity was observed in the dose-response relationship when these data were pooled and the dose was expressed as particle surface area per gram of lung (Appendix A). Based on the available data, the rat is the most sensitive rodent species to the pulmonary effects (inflammation and tumors) of inhaling PSLT particles such as TiO$_2$.

3.5.2 Comparison of Rodent and Human Lung Responses to PSLT Including TiO$_2$

3.5.2.1 Particle retention and lung response

The similarities and differences between human and rat lung structures, particle deposition, and responses to inhaled poorly soluble particles including TiO$_2$ have been described [Green et al. 2007; Baan 2007; Hext et al. 2005; Nikula et al. 2001; Heinrich 1996; Watson and Valberg 1996]. For example, there are structural differences in the small airways in the lungs of rats and humans (e.g., lack of well-defined respiratory bronchioles in rats). However, this region of the lungs is a primary site of particle deposition in both species, and particles that deposit in this region can translocate into the interstitium where they can elicit inflammatory and fibrotic responses [Green et al. 2007].

Both similarities and differences in rat and human nonneoplastic responses to inhaled particles were observed in a comparative pathology study (based on lung tissues from autopsy studies in workers and in chronic inhalation studies in rats, with chronic exposure to either coal dust, silica, or talc in both species) [Green et al. 2007]. Humans and rats showed some consistency in response by type of dust; i.e., granulomatous nodular inflammation was more severe among workers and rats exposed to silica or talc compared to coal dust. Similarly, humans and rats showed some graded response by dose; e.g., more severe centriacinar fibrosis at high versus low coal dust exposure in both humans and rats. In humans, the centriacinar fibrotic response was more severe among individuals exposed to either silica or coal dust compared to the rat response to these dusts. In rats, intra-alveolar acute inflammation, lipoproteinosis, and alveolar epithelial hyperplasia responses were more severe following chronic exposure to silica, talc, or coal dust compared to those responses in humans.

The greater inflammatory and cell proliferation responses in rats suggest they may be more susceptible than humans to lung tumor responses to inhaled particles via a secondary genotoxic mechanism involving chronic pulmonary inflammation, oxidative stress, and cell damage and proliferation. However, it is also important to note that alveolar epithelial hyperplasia was clearly observed in workers exposed to silica, talc, or coal dust (and this response was more severe in humans compared to rats at "low" coal dust exposure). An important consideration in interpreting these study results is that the quantitative dust exposures were well known in rats and poorly known in humans [Green et al. 2007]. The rat exposures to these dusts ranged from 2 to 18 mg/m^3 for 2 years. In humans, the exposure concentrations and durations were not reported for silica and talc; for coal dust, the durations were not reported, and "high" versus "low" exposure was defined as having worked before or after the 2 mg/m^3 standard. Since it is not possible from these data to determine how well the dust exposures and lung burdens compare between the rats and humans, the qualitative comparisons are probably more reliable than the quantitative comparisons, which could be due in part to unknown differences in the lung dose in rats and humans.

Different patterns of particle retention have been observed in rats, monkeys, and humans [Nikula et al. 2001]. Although no data are available on TiO_2 particle retention patterns, coal dust (fine size) or diesel exhaust particles (ultrafine) were retained at a higher volume percentage in the alveolar lumen in rats and in the interstitium in monkeys [Nikula et al. 1997]. The animals had been exposed by inhalation to 2 mg/m^3 of coal dust and/or diesel exhaust particulate for 2 years [Lewis et al. 1989]. A greater proportion of particles was also retained in the interstitium in humans compared to rats [Nikula et al. 2001]. In humans, the proportion of particles in the interstitium increased as the duration of exposure and estimated coal dust concentration increased. In rats, the particle retention pattern did not vary with increasing concentration of diesel exhaust particulate from 0.35 to 7.0 mg/m^3. The authors [Nikula et al. 1997, 2001] suggest that the difference in dust retention patterns may explain the apparent greater susceptibility of the rat to develop particle-related lung cancer compared to other species. However, these studies did not examine the particle retention patterns in rats at overloading doses, which are the doses at which elevated lung tumors are observed. During overload, the particle lung burden increases due to greater retention or sequestration of particles in the alveolar lumen and increased translocation of particles to the lung interstitium and lymph nodes [Lauweryns and Baert 1977; Adamson and Bowden 1981; Morrow 1988; Bellmann et al. 1991; Muhle et al. 1991; Oberdörster et al. 1992; Adamson and Hedgecock 1995]. The increased interstitialization of particles in rats at high doses has also been inferred by the increasing particle mass measured in the lung-associated lymph nodes [Bermudez et al. 2002] since the movement of particles from the alveolar region to the pulmonary lymphatics requires transport through the alveolar epithelium and its basement membrane [Lauweryns and Baert 1977]. Thus, the particle retention observed in rats at overloading doses may better represent the particle retention in the lungs of workers in dusty jobs, such as coal miners, where little or no particle clearance was observed among retired miners [Freedman and Robinson 1998]. In addition, similar particle retention patterns (i.e., location in lungs) have been observed in rats, mice, and hamsters [Bermudez et al. 2002], yet the rat lung response is more severe. The retained dose is clearly a main factor in the lung response, and the low response in the hamster has been attributed to its fast clearance and low retention of particles [Bermudez et al. 2002]. These findings suggest that the particle retention pattern is not the only, or necessarily the most important, factor influencing lung response between species including rat and human.

Particle size and ability to translocate from the lung alveolar region into the interstitium may also influence lung tissue responses. Borm et al. [2000] noted that the rat lung tumor response increased linearly with chronic pulmonary inflammation (following IT of various types of fine-sized particles), but that the rat lung tumor response to ultrafine TiO_2 was much greater relative to the inflammation response. They suggested that tumor development from ultrafine particles may be due to high interstitialization rather than to overload and its sequella as seen for the fine-sized particles. Although this hypothesis remains unproven, it is clear that the rat lung tumor response to ultrafine TiO_2 was greater than that of fine TiO_2 on a mass basis [Lee et al. 1985; Heinrich et al. 1995]. The rat lung tumor response was consistent across particle size on a particle surface area basis, suggesting that the particle surface is key to eliciting the response.

The extent of particle disaggregation in the lungs could influence the available particle surface as well as the ability of particles to translocate from the alveolar lumen into the interstitium, each of which could also influence pulmonary responses, as shown with various sizes of ultrafine and fine particles such as TiO_2 [Oberdörster 1996]. In a study using simulated lung surfactant, disaggregation of P25 ultrafine TiO_2 was not observed [Maier et al. 2006]. Another nanoscale TiO_2 particle (4 nm primary particle diameter; 22 nm count median diameter; 330 m^2/g specific surface area) was observed inside cells and cell organelles including the nucleus [Geiser et al. 2005]. Based on this observation, it has been suggested that an alternative genotoxic mechanism for nanoscale particles might involve direct interaction with DNA [Schins and Knaapen 2007]. NIOSH is not aware of any studies of the carcinogenicity of discrete nanoscale TiO_2 particles that would provide information to test a hypothesis of direct genotoxicity. However, this possible mechanism has been considered in the risk assessment (Chapters 4 and 5).

3.5.2.2 Inflammation in rats and humans

In rats that have already developed particle-associated lung tumors, the percentage of PMNs in the lungs is relatively high (e.g., 40%–60%) [Muhle et al. 1991]. However, it is not known what sustained level of PMNs is required to trigger epithelial proliferation and tumorigenic responses. In a chronic inhalation study, an average level of approximately 4% PMNs in BALF was measured in rats at the first dose and time point that was associated with a statistically significantly increased lung clearance half-time (i.e., a relatively early stage of overloading) [Muhle et al. 1991]. Similarly, in a subchronic inhalation study, 4% PMNs was predicted in rats at an average lung dose of TiO_2 or $BaSO_4$ that was not yet overloaded (based on measured particle lung burden) [Tran et al. 1999]. The relatively slight differences in the rat responses in these studies [Muhle et al. 1991; Tran et al. 1999] may be due in part to differences in rat strains (Fisher-344; Wistar) and measures to assess overloading (clearance half-time and lung burden retention).

In humans, chronic inflammation has been associated with nonneoplastic lung diseases in workers with dusty jobs. Rom [1991] found a statistically significant increase in the percentage of PMNs in BALF of workers with respiratory impairment who had been exposed to asbestos, coal, or silica (4.5% PMN in cases vs. 1.5% PMNs in controls). Elevated levels of PMNs have been observed in the BALF of miners with simple coal workers' pneumoconiosis (31% of total BALF cells vs. 3% in controls) [Vallyathan et al. 2000] and in patients with acute silicosis (also a ten-fold increase over controls) [Lapp and Castranova 1993; Goodman et al. 1992]. Humans with lung diseases that are characterized by chronic inflammation and epithelial cell proliferation (e.g., idiopathic pulmonary fibrosis, diffuse interstitial fibrosis associated with pneumoconiosis) have an increased risk of lung cancer [Katabami et al. 2000]. Dose-related increases in lung cancer have been observed in workers exposed to respirable crystalline silica [Rice et al. 2001; Attfield and Costello 2004], which can cause inflammation and oxidative tissue damage [Castranova 2000]. Chronic inflammation appears to be important in the etiology of dust-related lung disease, not only in rats, but also in humans with dusty jobs [Castranova 1998, 2000]. Case studies of nonmalignant lung disease in TiO_2 workers have reported lung responses indicative of inflammation, including pulmonary alveolar proteinosis [Keller et al. 1995] and interstitial fibrosis [Yamadori et al. 1986; Moran et al. 1991; Elo et al. 1972].

The percentage of PMNs in BALF of normal controls "rarely exceeded" 4% [Haslam et al. 1987]. In other human studies, "normal" PMN percentages ranged from 2%–17% [Costabel et al. 1990]. An average of 3% PMNs (range 0%–10%) was observed in the BALF of nonsmoker controls (without lung disease), while 0.8% (range 0.2%–3%) was seen in smoker controls (yet the PMN count in smokers was higher; this was due to elevation in other lung cell populations in smokers) [Hughes and Haslam 1990]. An elevation of 10% PMNs in the BALF of an individual has also been cited as being clinically abnormal [Martin et al. 1985; Crystal et al. 1981].

3.5.2.3 Noncancer responses to TiO_2 in rats and humans

Case studies of workers exposed to TiO_2 provide some limited information to compare the human lung responses to those observed in the rat. In both human and animal studies, TiO_2 has been shown to persist in the lungs. In some workers, extensive pulmonary deposition was observed years after workplace exposure to TiO$_2$ had ceased [Määttä and Arstila 1975; Rode et al. 1981]. This suggests that human lung retention of TiO$_2$ may be more similar to that in rats and mice compared to hamsters, which continue to clear the particles effectively at exposure concentrations that caused overloading in rats and mice [Everitt et al. 2000; Bermudez et al. 2002, 2004].

Inflammation, observed in pathologic examination of lung tissue, was associated with titanium (by X-ray or elemental analysis) in the majority of human cases with heavy TiO$_2$ deposition in the lung [Elo et al. 1972; Rode et al. 1981; Yamadori et al. 1986; Moran et al. 1991]. Pulmonary inflammation has also been observed in studies in rats, mice, and hamsters exposed to TiO$_2$ [Lee et al. 1985, 1986a; Everitt et al. 2000; Bermudez et al. 2002]. Continued pulmonary inflammation in the lungs of some exposed workers after exposure cessation [Määttä and Arstila 1975; Rode et al. 1981] also appears to be more consistent with the findings in rats and mice than in hamsters, where inflammation gradually resolved with cessation of exposure.

In one case study, pulmonary alveolar proteinosis (lipoproteinosis) was reported in a painter whose lung concentrations of titanium (60–129 × 10^6 particles/cm^3 lung tissue) were "among the highest recorded" in a database of similar analyses [Keller et al. 1995]. The titanium-containing particles were "consistent with titanium dioxide" and were the major type of particle found in the workers' lungs. In this case, the lipoproteinosis appeared more extensive than the lipoproteinosis seen in TiO$_2$-exposed rats [Lee et al. 1985, 1986a; Bermudez et al. 2002]. Thus, although the rat lipoproteinosis response was more severe than that in workers exposed to various fine-sized dusts [Green et al. 2007], this case study illustrates that humans can also experience severe lipoproteinosis response which is associated with particle retention in the lungs.

Mild fibrosis has been observed in the lungs of workers exposed to TiO$_2$ [Elo et al. 1972; Moran et al. 1991; Yamadori et al. 1986] and in rats with chronic inhalation exposure to TiO$_2$ [Heinrich et al. 1995; Lee et al. 1985, 1986a]. Alveolar metaplasia has been described in three human patients who had inhalation exposures to airborne TiO$_2$ [Moran et al. 1991]. In laboratory animals, alveolar metaplasia has been observed in rats, but not mice or hamsters, with subchronic inhalation to fine and ultrafine TiO$_2$ [Lee et al. 1985; Everitt et al. 2000; Bermudez et al. 2002, 2004].

Although these studies suggest some similarities in the lung responses reported in case studies of workers exposed to respirable TiO$_2$ and in experimental studies in mice and rats, the human studies are limited by being observational in nature and by lacking quantitative exposure data. Information was typically not available on various factors including other exposures that could have contributed to these lung responses in the workers. Also, systematic histopathological comparisons were not performed, for example, on the specific alveolar metaplastic changes of the rat and human lungs.

3.5.2.4 Lung tumor types in rats and humans

Lung tumors observed in rats following chronic inhalation of TiO$_2$ include squamous cell keratinizing cysts, bronchioloalveolar adenomas, squamous cell carcinomas, and adenocarcinomas [Heinrich et al. 1995; Lee et al. 1985] (see Section 3.2.5). The significance of the squamous cell keratinizing cystic tumor (a.k.a. proliferative keratin cyst) for human risk assessment has been evaluated [Carlton 1994; Boorman et al. 1996]. Squamous cell keratinizing cystic tumors are most prevalent in female rats exposed to high mass or surface area concentrations of PSLT. In a recent reanalysis of the lung tumors in the Lee et al. [1985] chronic inhalation study of fine (pigment-grade) TiO_2, the 15 lesions originally recorded as "squamous cell carcinoma" were reclassified as 16 tumors including two squamous metaplasia, 13 pulmonary keratin cysts, and one squamous cell carcinoma [Warheit and Frame 2006]. This reevaluation is consistent with the earlier evaluation of the squamous cell keratinizing cystic tumors by Boorman et al. [1996]. The classification of the 29 bronchioloalveolar adenomas in Warheit and Frame [2006] remained unchanged from that reported in Lee et al. [1985].

Human and rat lung cancer cell types have similarities and differences in their histopathologies [ILSI 2000; Maronpot et al. 2004]. The respiratory tracts of humans and rats are qualitatively similar in their major structures and functions, although there are also specific differences such as the absence of respiratory bronchioles in rats [Miller 1999; Green et al. 2007]. In humans, the major cell types of lung cancer worldwide are adenocarcinoma and squamous-cell carcinoma (also observed in rats) and small- and large-cell anaplastic carcinoma (not seen in rats) [Everitt and Preston 1999; ILSI 2000]. In rats exposed to PSLT, most cancers are adenocarcinomas or squamous-cell carcinomas of the alveolar ducts [ILSI 2000]. In recent years, there has been a shift in the worldwide prevalence of human lung tumors toward adenocarcinomas in the bronchoalveolar region [ILSI 2000; Maronpot et al. 2004; Kuschner 1995].

Maronpot et al. [2004] suggest that some of the apparent difference in the bronchioloalveolar carcinoma incidence in rodents and humans can be explained by differences in terminology, and that a more accurate comparison would be to combine the adenocarcinomas and bronchioloalveolar carcinomas in humans, which would significantly reduce the apparent difference. Cigarette smoking in humans is also likely to contribute to the difference between the incidences of human and rodent tumor types. Maronpot et al. [2004] suggest that if the smoking-related tumor types were eliminated from the comparison, then the major lung tumor types in humans would be adenocarcinomas and bronchioloalveolar carcinomas, which would correspond closely to the types of lung tumors occurring in rodents.

3.6 Rat Model in Risk Assessment of Inhaled Particles

The extent to which the rat model is relevant to predicting human lung doses and responses to inhaled particles including TiO_2 has been the subject of debate [Watson and Valberg 1996; ILSI 2000; Maronpot et al. 2004; Hext et al. 2005; Baan 2007]. Lung clearance of particles is slower in humans than in rats, by approximately an order of magnitude [Hseih and Yu 1998; Snipes 1996; ICRP 1994], and some humans (e.g., coal miners) may be exposed to concentrations resulting in doses that would overload particle clearance from rat lungs [Morrow et al. 1991; Freedman and Robinson 1988]. Thus, the doses that cause overloading in rats appear to be relevant to estimating disease risk in workers given these similarities in particle retention in the lungs of rats at overloading doses and in workers with high dust exposures.

Some have stated that the inhalation dose-response data from rats exposed to PSLT particles should not be used in extrapolating cancer risks to humans because the rat lung tumor response has been attributed as a rat-specific response to the overloading of particle clearance from the lungs [Levy 1996; Watson and Valberg 1996; Hext et al. 2005; Warheit and Frame 2006]. While some of the tumors (keratinizing cystic tumors) may not be relevant to humans, other types of tumors (adenomas, adenocarcinoma, squamous cell carcinoma) do occur in humans.

Mice and hamsters are known to give false negatives to a greater extent than rats in bioassays for some particulates that have been classified by the IARC as human carcinogens (limited or sufficient evidence), including crystalline silica and nickel subsulfide; and the mouse lung tumor response to other known human particulate carcinogens—including beryllium, cadmium, nickel oxide, tobacco smoke, asbestos, and diesel exhaust—is substantially less in mice than that in rats [Mauderly 1997]. These particulates may act by various plausible mechanisms, which are not fully understood [Schins and Knaapen 2007; Castranova 1998]. The risks of several known human particulate carcinogens would thus be underestimated by using dose-response data from rodent models other than the rat. Although the mechanism of particle-elicited lung tumors remains to be fully elucidated, the rat and human lung responses to poorly soluble particles of low or high toxicity (e.g., coal dust and silica) are qualitatively similar in many of the key steps for which there are data, including pulmonary inflammation, oxidative stress, and alveolar epithelial cell hyperplasia [Vallyathan et al. 1998; Castranova 2000; Green et al. 2007; Baan 2007]. Case studies of lung responses in humans exposed to TiO_2 are consistent in showing some similarities with rat responses (see Section 3.5.2.3). Semiquantitative comparisons of the lungs of rats and humans exposed to diesel exhaust, coal dust, silica, or talc indicate both similarities and differences, including similar regions of particle retention in the lungs but at different proportions [Nikula et al. 2001], and greater fibrotic response in humans but greater inflammation and epithelial hyperplasia responses in rats [Green et al. 2007]. These sensitive inflammatory and proliferative responses to particles in the lungs are considered key to the rat lung tumor response [ILSI 2000]. Although the data are limited for quantitative comparison of rat and human dose-response relationships to inhaled particles, the available data (e.g., crystalline silica and diesel exhaust particles) indicate that the rat-based estimates are not substantially greater, and some are lower, than the human-based risk estimates [Stayner et al. 1998; Kuempel et al. 2001c, 2002]. Collectively, these studies indicate that, while there are uncertainties about the rat as a model for particle-related lung cancer in humans, and specifically for TiO_2, there is insufficient evidence for concluding that the rat is not a valid model. Expert advisory panels have concluded that chronic inhalation studies in rats are the most appropriate tests for predicting the inhalation hazard and risk of fibers to humans [Vu et al. 1996] and that, in the absence of mechanistic data to the contrary, it is reasonable to assume that the rat model can identify potential carcinogenic hazards of poorly soluble particles to humans [ILSI 2000], including PSLT such as TiO_2.

4 Quantitative Risk Assessment

4.1 Data and Approach

Dose-response data are needed to quantify risks of workers exposed to TiO_2. Such data may be obtained either from human studies or extrapolated to humans from animal studies. The epidemiologic studies on lung cancer have not shown a dose-response relationship in TiO_2 workers [Fryzek et al. 2003; Boffetta et al. 2004]. However, dose-response data are available in rats, for both cancer (lung tumors) and early, noncancer (pulmonary inflammation) endpoints. The lung tumor data (see Table 4–4) are from chronic inhalation studies and include three dose groups for fine TiO_2 and one dose group for ultrafine TiO_2 (in addition to controls). The pulmonary inflammation data are from subchronic inhalation studies of fine and ultrafine particles [Tran et al. 1999; Cullen et al. 2002; Bermudez et al. 2002; Bermudez et al. 2004]. Various modeling approaches are used to fit these data and to estimate the risk of disease in workers exposed to TiO_2 for up to a 45-year working lifetime.

The modeling results from the rat dose-response data provide the quantitative basis for developing the RELs for TiO_2, while the mechanistic data from rodent and human studies (Chapter 3) provide scientific information on selecting the risk assessment models and methods. The practical aspects of mass-based aerosol sampling and analysis were also considered in the overall approach (i.e., the conversion between particle surface area for the rat dose-response relationships and mass for the human dose estimates and RELs). Figure 4–1 illustrates the risk assessment approach.

4.2 Methods

Dose-response modeling was used to estimate the retained particle burden in the lungs associated with lung tumors or pulmonary inflammation. Both maximum likelihood (MLE) and 95% lower CI estimates of the internal lung doses in rats were computed. Particle surface area was the dose metric used in these models because it has been shown to be a better predictor than particle mass of both cancer and noncancer responses in rats (Chapter 3). In the absence of quantitative data comparing rat and human lung responses to TiO_2, rat and human lung tissue are assumed to have equal sensitivity to an equal internal dose in units of particle surface area per unit of lung surface area [EPA 1994].

4.2.1 Particle Characteristics

Study-specific values of particle MMAD, geometric standard deviation (GSD), and specific surface area were used in the dosimetric modeling when available (see Tables 4–1 and 4–4). The Heinrich et al. [1995] study reported a specific surface area (48 ± 2 m^2/g ultrafine TiO_2) for the airborne particulate, as measured by the BET N_2 adsorption method. For

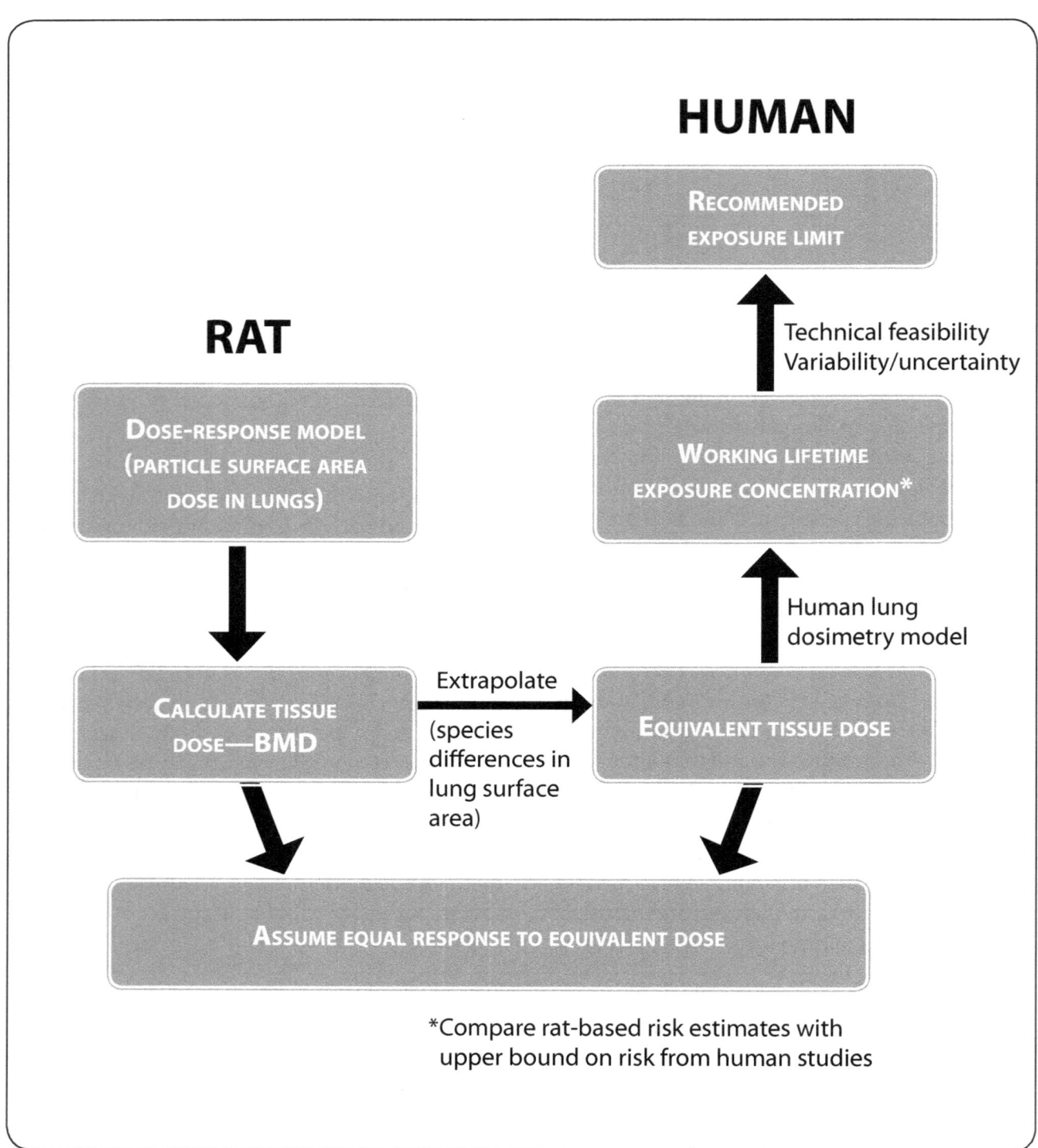

Figure 4–1. Risk assessment approach using rat dose-response data to derive recommended exposure limits for TiO_2.

the Lee et al. [1985] study, the specific surface area (4.99 m^2/g fine TiO_2) reported by Driscoll [1996] was used; that value was based on measurement of the specific surface area of a rutile TiO_2 sample similar to that used in the Lee study [Driscoll 2002]. This specific surface area was also assumed for the fine TiO_2 in the Muhle et al. [1991] study. Otherwise, fine TiO_2 was assumed to have the particle characteristics reported by Tran et al. [1999] and a specific surface area of 6.68 m^2/g, and ultrafine TiO_2 was assumed to have the particle characteristics reported by Heinrich et al. [1995] and a specific surface area of 48 m^2/g.

4.2.2 Critical Dose

The term "critical dose" is defined as the retained particle dose in the rat lung (maximum likelihood estimate [MLE] or 95% lower confidence limit [LCL]) associated with a specified response, including either initiation of inflammation or a given excess risk of lung cancer. One measure of critical dose is the *benchmark dose* (BMD), which has been defined as ". . . a statistical lower confidence limit on the dose corresponding to a small increase in effect over the background level" [Crump 1984]. In current practice, and as used in this document, the BMD refers to the MLE from the model; and the benchmark dose lower bound (BMDL) is the 95% LCL of the BMD [Gaylor et al. 1998], which is equivalent to the BMD as originally defined by Crump [1984]. For dichotomous noncancer responses the benchmark response level is typically set at a 5% or 10% excess risk; however, there is less agreement on benchmark response levels for continuous responses. The benchmark response level used in this analysis for pulmonary inflammation, as assessed by elevated levels of PMNs, was 4% PMNs in BALF [Tran et al. 1999] (see Section 3.5.2.2). Another measure of critical dose, used in an earlier draft of this current intelligence bulletin, was the estimated threshold dose derived from a piecewise linear model fit to the pulmonary inflammation data (Appendix B). As discussed in Section 4.3.1.2, not all of the current data sets are compatible with a threshold model for pulmonary inflammation; however, the threshold estimates described in Appendix B have been retained in this document for comparative purposes. For lung tumors the approach to estimating critical lung doses was to determine the doses associated with a specified level of excess risk (e.g., 1 excess case per 1,000 workers exposed over a 45-year working lifetime), either estimated directly from a selected model or by model averaging using a suite of models [Wheeler and Bailer 2007]. The 1 per 1,000 lifetime excess risk level was considered to be a significant risk based on the "benzene" decision [U.S. Supreme Court 1980].

4.2.3 Estimating Human Equivalent Exposure

The critical doses were derived using particle surface area per gram of rat lung, which was estimated from the mass lung burden data, rat lung weights, and measurements or estimates of specific surface area (i.e., particle surface area per mass). The use of a particle surface area per gram of lung dose for the rat dose-response analyses was necessary in order to normalize the rat lung weights to a common dose measure, since several strains of rats with varying lung weights were used in the analyses (see Table 4–1 and Table 4–4), and it would be improper to equate the total lung doses of rats with greatly varying lung weights (Table 4–4). The critical doses were then multiplied by 1.5 in order to normalize them to rats of the size used as a reference for lung surface area, as these were estimated to have lung weights of approximately 1.5 grams, based on the animal's body weights. The critical doses were then extrapolated to humans based on the ratio of rat lung to human lung surface areas, which were assumed to be 0.41 m² for Fischer 344 rats, 0.4 m² for Sprague-Dawley rats, and 102.2 m² for humans [Mercer et al. 1994]. These critical particle surface area doses were then converted back to particle mass dose for humans because the current human lung dosimetry models (used to estimate airborne concentration leading to the critical lung doses) are all mass-based and because the current occupational exposure limits for most airborne particulates including TiO_2 are also mass-based.

4.2.4 Particle Dosimetry Modeling

The multiple-path particle dosimetry model, version 2, (MPPD2) human lung dosimetry model [CIIT and RIVM 2002] was used to estimate the working lifetime airborne mass concentrations associated with the critical doses in human lungs, as extrapolated from the rat dose-response data. The specific MPPD2 module used was the Yeh/Schum Symmetric model, assuming 17.5 breaths per minute and a tidal volume of 1,143 ml, with exposures of 8 hours per day, 5 days per week for a working lifetime of 45 years. The total of alveolar TiO_2 plus TiO_2 in the lung-associated lymph nodes was considered to be the critical human lung dose.

The respiratory frequency and tidal volume were chosen to be consistent with the International Commission on Radiological Protection (ICRP) parameter values for occupational exposure, which equate an occupational exposure to 5.5 hours/day of light exercise and 2.5 hours/day of sedentary sitting, with a total inhalation volume of 9.6 m³ in an 8-hour day [ICRP 1994]. The ICRP assumes 20 breaths per minute and a tidal volume of 1,250 ml for light exercise and 12 breaths per minute and a tidal volume of 625 ml for sedentary sitting. The values assumed in this analysis for modeling TiO_2 exposures—i.e., 17.5 breaths per minute and a tidal volume of 1,143 ml—are a weighted average of the respiratory values for the light exercise and sedentary sitting conditions and are designed to match the ICRP [1994] value for total daily occupational inhalation volume.

In summary, the dose-response data in rats were used to determine the critical dose associated with pulmonary inflammation or lung tumors, as particle surface area per surface area of lung tissue. The working lifetime airborne mass concentrations associated with the human-equivalent critical lung burdens were estimated using human lung dosimetry models. The results of these quantitative analyses and the derivation of the RELs for fine and ultrafine TiO_2 are provided in the remainder of this chapter.

4.3 Dose-Response Modeling of Rat Data and Extrapolation to Humans

4.3.1 Pulmonary Inflammation

4.3.1.1 Rat data

Data from four different subchronic inhalation studies in rats were used to investigate the relationship between particle surface area dose and pulmonary inflammation response: (1) TiO_2 used as a control in a study of the toxicity of volcanic ash [Cullen et al. 2002], (2) fine TiO_2 and $BaSO_4$ in a study of the particle surface area as dose metric [Tran et al. 1999], (3) fine TiO_2 in a multidose study [Bermudez et al. 2002], and (4) ultrafine TiO_2 in a multidose study

[Bermudez et al. 2004]. Details of these studies are provided in Table 4–1. Individual rat data were obtained for PMN count in the lungs in each study. In the Tran et al. study and the two Bermudez et al. studies different groups of rats were used to measure lung burden and PMNs, and group average values were used to represent the lung burden of TiO_2 in each exposure group. In the Cullen et al. study, PMN and lung burden data were obtained for each individual rat. Data from the two Bermudez et al. studies were combined and treated as a single study for dose-response analysis purposes.

4.3.1.2 Critical dose estimates in rats

The TiO_2 pulmonary inflammation data from the Tran et al. [1999] and Cullen et al. [2002] studies could be fitted with a piecewise linear model which included a threshold parameter (described in Appendix B), and the threshold parameter estimate was significantly different from zero at a 95% confidence level. The MLE of the threshold dose was 0.0134 m² (particle surface area for either fine or ultrafine TiO_2) for TiO_2 alone (90% CI = 0.0109–0.0145) based on data from Tran et al. [1999] and 0.0409 m² (90% CI = 0.0395–0.0484) based on data from Cullen et al. [2002]. However, the fine and ultrafine TiO_2 pulmonary inflammation data from the Bermudez et al. [2002] and Bermudez et al. [2004] data sets provided no indication of a nonzero response threshold and were not consistent with a threshold model. Therefore, critical dose estimation for the pulmonary inflammation data was carried out via a benchmark dose approach, since benchmark doses could be fit to all three of the data sets (Tran et al. [1999], Cullen et al. [2002], and the combined data from Bermudez et al. [2002] and Bermudez et al. [2004]).

Continuous models in the benchmark dose software (BMDS) suite [EPA 2007] were fitted to the pulmonary inflammation data using percent neutrophils as the response and TiO_2 surface area (m²) per gram of lung as the predictor. The model that fit best for all three sets of TiO_2 data was the Hill model, which converged and provided an adequate fit. Since there appears to be an upper limit to the degree of physiological response, the Hill model is able to capture this behavior better than the linear, quadratic, or power models. Models were fitted using constant variance for the Cullen et al. [2002] data and a nonconstant variance for the Tran et al. [1999] data and the combined data from Bermudez et al. [2002] and Bermudez et al. [2004]. The model fits for these data sets are illustrated in Figures 4-2, 4-3, and 4-4, respectively. In all models the critical dose or BMD was defined as the particle surface area per gram of lung tissue associated with a 4% inflammatory response of neutrophils, which has been equated to a low level inflammatory response [Tran et al. 1999]; see Section 3.5.2.2. The benchmark dose estimates for pulmonary inflammation in rats are shown in Table 4–2.

4.3.1.3 Estimated human equivalent exposure

The critical dose estimates from Table 4–2 were converted to mass dose and extrapolated to humans by adjusting for species differences in lung surface area, as described in Section 4.2.3. Pulmonary dosimetry modeling was used to estimate the occupational exposure concentrations corresponding to the benchmark dose estimates, as described in Section 4.2.4. Lifetime occupational exposure concentrations estimated to produce human lung burdens equivalent to the rat critical dose levels for inflammation are presented in Table 4–3,

Table 4–1. Comparison of rat inhalation studies used to model the relationship between TiO$_2$ and pulmonary inflammation

Experimental conditions	Study			
	Tran et al. [1999]	Cullen et al. [2002]	Bermudez et al. [2002]	Bermudez et al. [2004]
TiO$_2$ particle size: MMAD (GSD)	2.1 (2.2) μm	1.2 (2.2) μm	1.44 (1.71) μm	1.44 (2.60) μm (agglomerated)
Specific surface area	6.68 m^2/g	6.41 m^2/g	6.68 m^2/g (estimated)	48 m^2/g (estimated)
Sex, rat strain	Male, Wistar rats	Male, Wistar rats	Female, Fischer 344 rats	Female, Fischer 344 rats
Exposure conditions	Whole body inhalation	Nose-only inhalation	Whole body inhalation	Whole body inhalation
	7 hr/day, 5 days/week	6 hr/day, 5 days/week	6 hr/day, 5 days/week	6 hr/day, 5 days/week
TiO$_2$ concentration, duration	25 mg/m^3, 7.5 months 50 mg/m^3, 4 months	140 mg/m^3, 2 months	10, 50, or 250 g/m^3, 13 weeks	0.5, 2, or 10 mg/m^3, 13 weeks

MMAD = mass median aerodynamic diameter; GSD = geometric standard deviation

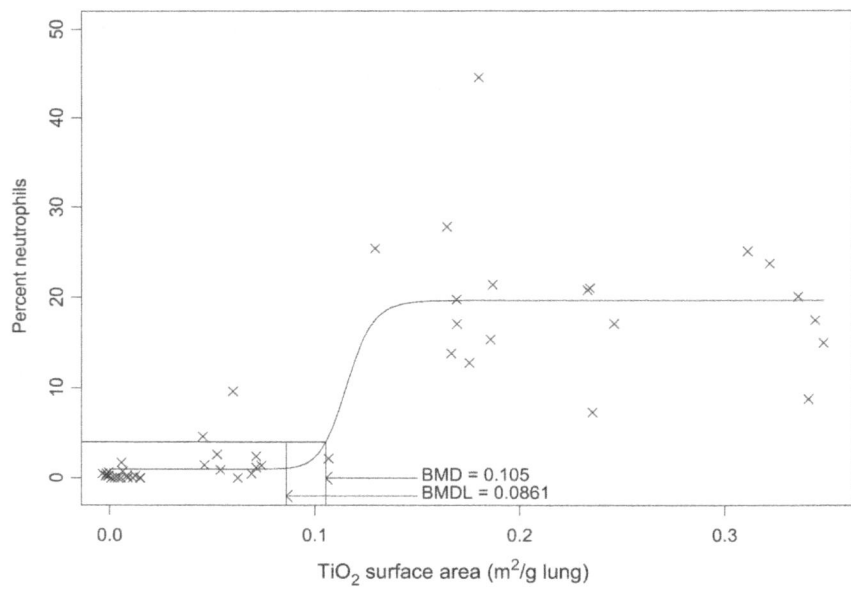

Figure 4–2. Hill model fit to rat data on pulmonary inflammation (percent neutrophils) and particle surface area dose of TiO$_2$

Data source: Cullen et al. [2002]

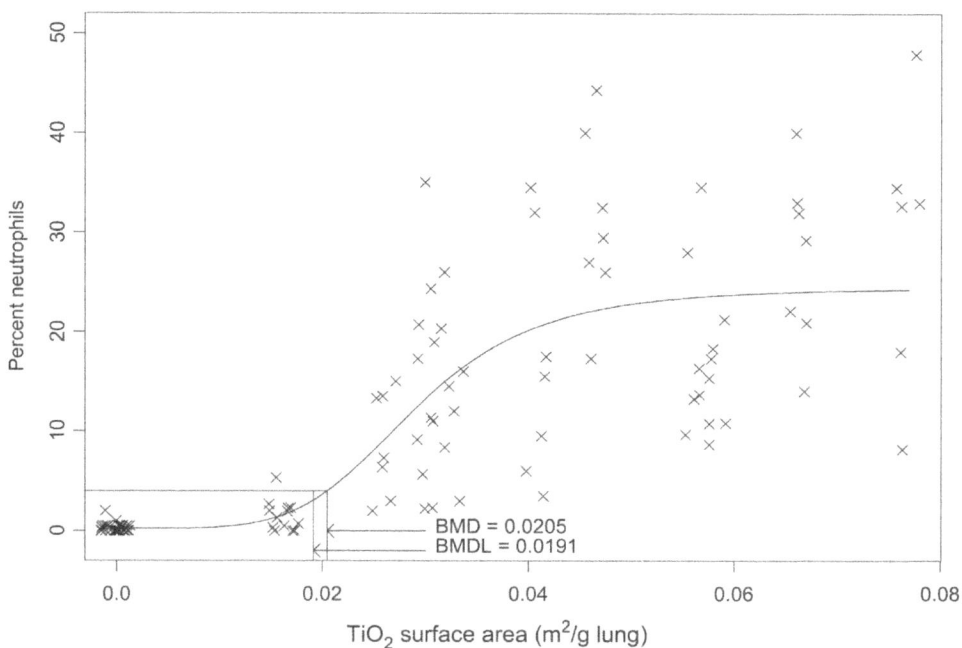

Figure 4–3. Hill model fit to rat data on pulmonary inflammation (percent neutrophils) and particle surface area dose of TiO_2

Data source: Tran et al. [1999]

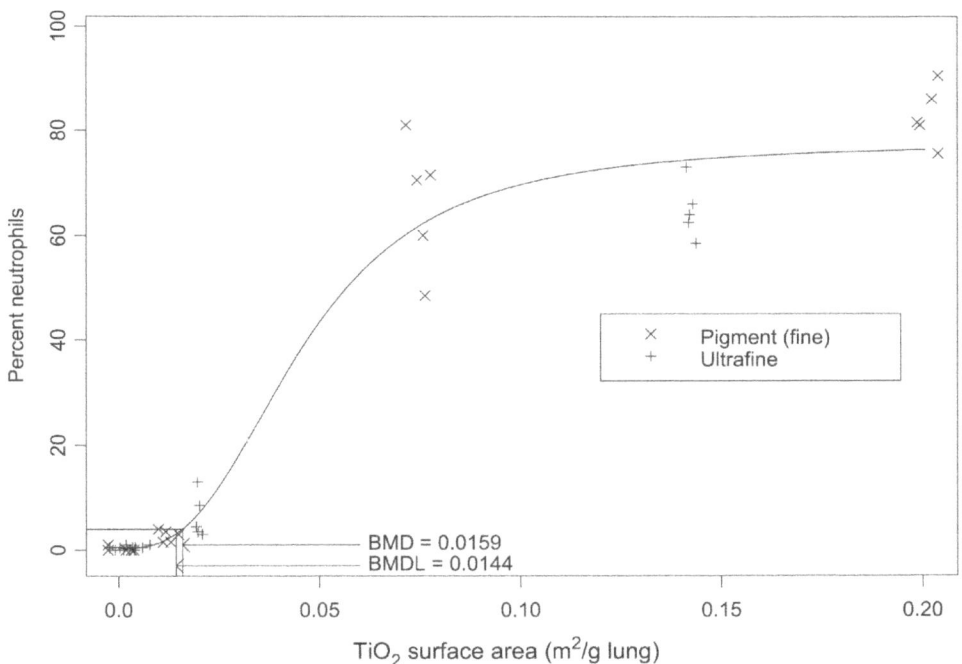

Figure 4–4. Hill model fit to rat data on pulmonary inflammation (percent neutrophils) and particle surface area dose of TiO_2

Data source: (combined) Bermudez et al. [2002], Bermudez et al. [2004]

Table 4–2. Benchmark dose estimates for particle surface area dose (m²) per gram of lung associated with pulmonary inflammation in rats (as PMNs in BALF), based on a Hill model

Data modeled	MLE	95% LCL
TiO_2 [Tran et al. 1999]	0.0205	0.0191
TiO_2 [Cullen et al. 2002]	0.1054	0.0861
TiO_2 [Bermudez et al. 2002 and Bermudez et al. 2004, combined]	0.0159	0.0144

BALF = bronchoalveolar lavage fluid; LCL = lower confidence limit; MLE = maximum likelihood estimate; PMNs = polymorphonuclear leukocytes; TiO_2 = titanium dioxide.

Table 4–3. Estimated mean airborne mass concentrations of fine and ultrafine TiO_2 in humans and related human lung burdens (TiO_2 surface area dose) associated with pulmonary inflammation after a 45-year working lifetime

	Critical dose in human lungs*					
	Particle surface area (m²/lung)		Particle mass (g/lung)		MPPD (ICRP) lung model (mg/m³)†	
Particle size and study	MLE	95% LCL	MLE	95% LCL	MLE	95% LCL
Fine TiO_2 (2.1 μm, 2.2 GSD; 6.68 m²/g):						
Tran et al. [1999]	7.86	7.32	1.18	1.10	1.11	1.03
Cullen et al. [2002]	40.39	33.00	6.30	5.15	5.94	4.86
Bermudez et al. [2002 and 2004]	6.09	5.52	0.91	0.83	0.86	0.78
Ultrafine TiO_2 (0.8 μm,§ 1.8 GSD; 48 m²/g):						
Tran et al. [1999]	7.86	7.32	0.164	0.153	0.136	0.127
Cullen et al. [2002]	40.39	33.00	0.842	0.687	0.698	0.570
Bermudez et al. [2002 and 2004]	6.09	5.52	0.127	0.115	0.105	0.095

MPPD = multiple-path particle dosimetry [CIIT and RIVM 2002] model, including ICRP [1994] clearance model; GSD = geometric standard deviation; ICRP = International Commission on Radiological Protection; LCL = lower confidence limit; MLE = maximum likelihood estimate; TiO_2 = titanium dioxide.

*MLE and 95% LCL were determined in rats (Table 4–2) and extrapolated to humans based on species differences in lung surface area, as described in Section 4.2.3.

†Mean concentration estimates derived from the CIIT and RIVM [2002] lung model, which includes the ICRP [1994] clearance model.

§Mass median aerodynamic diameter (MMAD). Ultrafine particle size is for agglomerate [Heinrich et al. 1995].

ranging from 0.78–5.94 mg/m³ for fine TiO_2 and 0.095–0.698 mg/m³ for ultrafine TiO_2. These exposure concentrations are presented only for comparison to the concentrations associated with lung tumors presented in Table 4–6. As discussed in Section 4.5.1, the inflammatory levels presented in Table 4–3 are estimates of frank-effect levels (as opposed to no-effect levels) and would normally be adjusted by animal-to-human and interindividual uncertainty factors before being used as the basis for a REL.

4.3.2 Lung Tumors

4.3.2.1 Rat data

Dose-response data from chronic inhalation studies in rats exposed to TiO_2 were used to estimate working lifetime exposures and lung cancer risks in humans. These studies are described in more detail in Table 4–4 and include fine (pigment-grade) rutile TiO_2 [Lee et al. 1985; Muhle et al. 1991] and ultrafine anatase TiO_2 [Heinrich et al. 1995]. The doses for fine TiO_2 were 5 mg/m³ [Muhle et al. 1991] and 10, 50, and 250 mg/m³ [Lee et al. 1985]. For ultrafine TiO_2 there was a single dose of approximately 10 mg/m³ [Heinrich et al. 1995]. Each of these studies reported the retained particle mass lung burdens in the rats. The internal dose measure of particle burden at 24 months of exposure was used in the dose-response models, either as particle mass or particle surface area (calculated from the reported or estimated particle surface area per gram).

The relationship between particle surface area dose of either fine or ultrafine TiO_2 and lung tumor response (including all tumors or tumors excluding the squamous cell keratinizing cysts) in male and female rats was shown in Chapter 3. Statistically significant increases in lung tumors were observed at the highest dose of fine TiO_2 (250 mg/m³) or ultrafine TiO_2 (approximately 10 mg/m³), whether or not the squamous cell keratinizing cysts were included in the tumor counts.

Different strains and sexes of rats were used in each of these three TiO_2 studies. The Lee et al. [1985] study used male and female Sprague-Dawley rats (crl:CD strain). The Heinrich study used female Wistar rats [crl:(WI)BR strain]. The Muhle et al. [1991] study used male and female Fischer-344 rats but reported only the average of the male and female lung tumor proportions. The body weights and lung weights differed by rat strain and sex (Table 4–4). These lung mass differences were taken into account when calculating the internal doses, either as mass (mg TiO_2/g lung tissue) or surface area (m² TiO_2/m² lung tissue).

4.3.2.2 Critical dose estimates in rats

Statistical models for quantal response were fitted to the rat tumor data, including the suite of models in the BMDS [EPA 2003]. The response variable used was either all lung tumors or tumors excluding squamous cell keratinizing cystic tumors. Figure 4–5 shows the fit of the various BMD models [EPA 2003] to the lung tumor response data (without squamous cell keratinizing cysts) in male and female rats chronically exposed to fine or ultrafine TiO_2 [Lee et al. 1985; Heinrich et al. 1995; and Muhle et al. 1991].

The lung tumor response in male and female rats was significantly different for "all tumors" but not when squamous cell keratinizing cystic tumors were removed from the analysis (Appendix A, Table A–2). In other words, the male and female rat lung tumor responses were equivalent except for the squamous cell

Table 4–4. Summary of chronic inhalation studies in rats exposed to TiO_2

Particle size and type; study	Rat strain	Mean body weight of controls at 24 months (g)		Mean lung weight of controls at 24 months (g)		Particle size MMAD (µm) and specific SA (m^2/g TiO_2)	Exposure concentration (mg/m^3)	Treated rats				
								Retained mean dose (mg TiO_2/lung)*		Tumor proportion (rats with tumors/total rats)		
		Female	Male	Female	Male			Female	Male	Female	Male	Average
Fine TiO_2 (≥ 99% *rutile*):												
Lee et al. [1985, 1986a]	Sprague-Dawley (crl: CD)	557	780	2.35	3.25	MMAD: 1.5 to 1.7 SA: 4.99 [Driscoll 1996]	0 10 50 250	0 32.3 130 545.8	0 20.7 118.3 784.8	0/77 0/75 0/74 14/74	2/79 2/71 1/75 12/77[†]	— — — —
Muhle et al. [1989, 1991, 1994]; Bellman et al. [1991]	Fischer-344	337	403	1.05	1.38	MMAD: 1.1 (GSD: 1.6) SA: 4.99 (estimate)	0 5	0 2.72	— —	— —	— —	3/100 2/100[§]

(Continued)

See footnotes at end of table.

Table 4–4 (Continued). Summary of chronic inhalation studies in rats exposed to TiO$_2$

Particle size and type; study	Rat strain	Mean body weight of controls at 24 months (g) Female	Mean body weight of controls at 24 months (g) Male	Mean lung weight of controls at 24 months (g) Female	Mean lung weight of controls at 24 months (g) Male	Particle size MMAD (μm) and specific SA (m^2/g TiO$_2$)	Exposure concentration (mg/m^3)	Treated rats Retained mean dose (mg TiO$_2$/lung)* Female	Treated rats Retained mean dose (mg TiO$_2$/lung)* Male	Tumor proportion (rats with tumors/total rats) Female	Tumor proportion (rats with tumors/total rats) Male	Tumor proportion (rats with tumors/total rats) Average
Ultrafine TiO$_2$ (~80% anatase; ~20% rutile):												
Heinrich et al. [1995]; Muhle et al. [1994]	Wistar [crl:(WI)BR]	417	—	1.44		MMAD: 0.80 (GSD: 1.8) (agglomerates)	0	0		At 24 months: 0/10 (controls) 4/9 (all tumors)		
						0.015–0.040 (individual particles) SA: 48 (SD: 2)	~10	39.29 (SD: 7.36)		At 30 months: 1/217 (controls) 19/100 (no keratinizing cysts) 32/100 (all tumors)**		

GSD = geometric standard deviation; MMAD = mass median aerodynamic diameter; SA = surface area (mean or assumed mean); SD = arithmetic standard deviation; TiO$_2$ = titanium dioxide; crl:CD and crl:(WI)BR are the rat strain names from Charles River Laboratories, Inc.

*Lung particle burdens in controls not reported; assumed to be zero.

†Tumor types: controls, male: 2 bronchioloalveolar adenomas. At 10 mg/m^3, males: 1 large cell anaplastic carcinoma and 1 bronchioloalveolar adenoma. At 50 mg/m^3, male: 1 bronchioloalveolar adenoma. At 250 mg/m^3, females: 13 bronchioloalveolar adenomas and 1 squamous cell carcinoma; males: 12 bronchioloalveolar adenomas. In addition to the tumors listed above, 13 keratin cysts were observed: 1 in a 10 mg/m^3 male, 1 in a 250 mg/m^3 male, and 11 in 250 mg/m^3 females.

§Dose was averaged for male and female rats because the tumor rates were reported only for male and female rats combined. Tumor types: controls, 2 adenocarcinomas and 1 adenoma. At 5 mg/m^3: 1 adenocarcinoma and 1 adenoma.

**Tumor types: controls, at 30 months: 1 adenocarcinoma. At ~10 mg/m^3: 20 benign squamous-cell tumors, 3 squamous-cell carcinomas, 4 adenomas, and 13 adenocarcinomas (includes 8 rats with 2 tumors each).

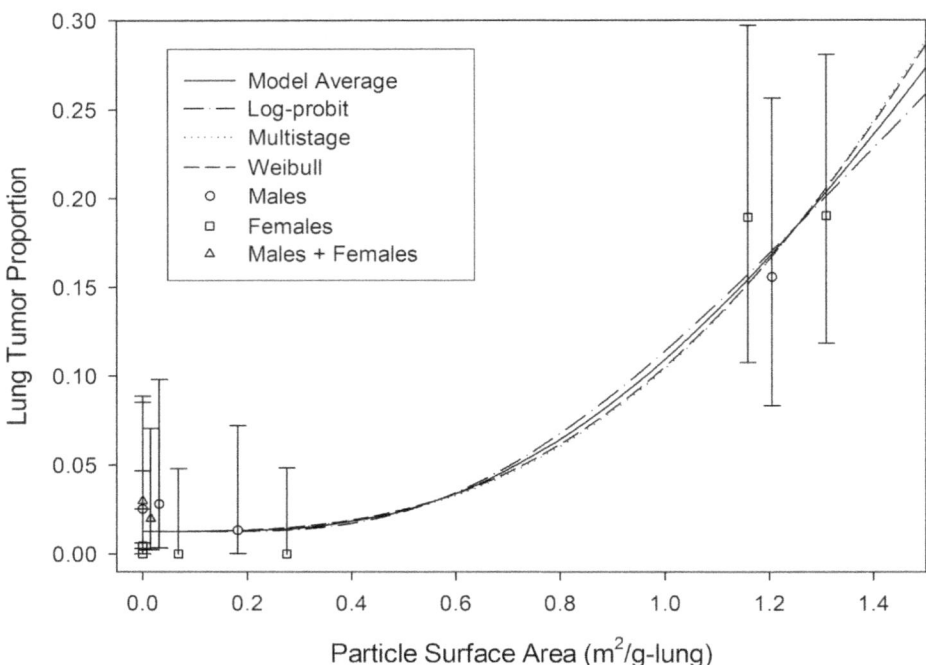

Figure 4–5. BMD models and three-model average fit to the lung tumor data (without squamous cell keratinizing cysts) in male and female rats chronically exposed to fine or ultrafine TiO_2

Note: The specific models used in constructing the model average were the multistage, Weibull, and log-probit models. The confidence intervals represent 95% binomial confidence limits for the individual data points.

Data source: BMD models and three-model average [Wheeler and Bailer 2007], male and female rats exposed to TiO_2 [Lee et al. 1985; Heinrich et al. 1995; Muhle et al. 1991]

keratinizing cystic tumor response, which was elevated only in the female rats. To account for the heterogeneity in the "all tumor" response among male and female rats [Lee et al. 1985; Heinrich et al. 1995], a modified logistic regression model was developed (Appendix A); this model also adjusted for the combined mean tumor response for male and female rats reported by Muhle et al. [1991]. As discussed in Chapter 3, many pathologists consider the rat lung squamous cell keratinizing cystic tumor to be irrelevant to human lung pathology. Excess risk estimates of lung tumors were estimated both ways—either with or without the squamous cell keratinizing cystic tumor data. The full results of the analyses including squamous cell keratinizing cystic tumors can be found in Appendix A. Inclusion of the keratinizing cystic tumors in the analyses resulted in slightly higher excess risk estimates in females, but not males. Since the male and female rat tumor responses may be combined when the squamous cell keratinizing cystic tumors are excluded and exclusion does not have a major numeric impact on the risk estimates, risk estimates for TiO_2-induced lung tumors are based on the combined male and female rat lung tumors, excluding the squamous cell keratinizing cystic tumors. All lung tumor-based risk estimates shown below have been derived on this basis. Classification of the squamous cell keratinizing cystic tumors in the Lee et al. [1985] study was based on a reanalysis of these lesions by Warheit and Frame [2006].

The estimated particle surface area dose associated with a 1/1000 excess risk of lung tumors is shown in Table 4–5 for lung tumors excluding squamous cell keratinizing cystic lesions. The estimated particle surface area doses—BMD and BMDL—associated with a 1/1000 excess risk of lung cancer vary considerably depending on the shape of the model in the low-dose region. The model-based estimates were then summarized using a model averaging technique [Wheeler and Bailer 2007], which weights the various models based on the model fit. Model averaging provides an approach for summarizing the risk estimates from the various models, which differ in the low-dose region which is of interest for human health risk estimation, and also provides an approach for addressing the uncertainty in choice of model in the BMD approach. The specific model averaging method used was the three-model average procedure as described by Wheeler and Bailer [2007], using the multistage, Weibull, and log-probit models. These received weights of 0.14, 0.382 and 0.478, respectively, in the averaging procedure. This approach was considered to be appropriate for the TiO_2 data set because the dose-response relationship is nonlinear, and the specific models used in the three-model average procedure do not impose low-dose linearity on the model average if linearity is not indicated by the data. In this case the best fitting models are all strongly sub-linear; the multistage model is cubic in form, the Weibull model is similar with a power of 2.94, and the log-probit model has a slope of 1.45. Therefore the dose-response relationship is quite steep, and the estimated risk drops sharply as the dose is reduced, as shown in Table 4–7.

4.3.2.3 Estimated human equivalent exposure

The critical dose estimates from Table 4–5 were converted to mass dose and extrapolated to humans by adjusting for species differences in lung surface area, as described in Section 4.2.3. Pulmonary dosimetry modeling was used to estimate the occupational exposure concentrations corresponding to the benchmark dose estimates, as described in Section 4.2.4. Lifetime occupational exposure concentrations estimated to produce human lung burdens equivalent to the rat critical dose levels for lung tumors are presented in Table 4–6 for fine and ultrafine TiO_2. Estimated occupational exposure levels corresponding to various levels of lifetime excess risk are shown in Table 4–7. This table shows the risk estimates for both fine and ultrafine TiO_2. The 95% LCL estimates of the occupational exposure concentrations expected to produce a given level of lifetime excess risk are shown in the right-hand column. The concentrations shown in bold for fine and ultrafine TiO_2 represent 1 per 1000 risk levels, which NIOSH has used as the basis for establishing RELs. The REL for ultrafine TiO_2 was rounded from 0.29 mg/m³ to 0.3.

4.3.3 Alternate Models and Assumptions

The choice of dosimetry models influences the estimates of the mean airborne concentration. A possible alternative to the MPPD model of

Table 4–5. BMD and BMDL estimates of TiO$_2$ particle surface area dose in rat lungs (m^2/g-lung) associated with 1/1000 excess risk of lung cancer*

Model: BMDS [EPA 2003]	P-value (for lack of fit)†	BMD	BMDL
Gamma	0.33	0.258	0.043
Logistic	0.31	0.035	0.027
Log-logistic‡	0.17	0.223	0.026
Log-probit‡	0.28	0.286	0.051
Multistage	0.40	0.218	0.016
Probit	0.29	0.030	0.022
Quantal-linear	0.15	0.0075	0.0058
Quantal-quadratic	0.38	0.092	0.081
Weibull	0.32	0.212	0.036
MA**	—	0.244	0.029

MA = model average; BMD = maximum-likelihood estimate of benchmark dose; BMDL = benchmark dose lower bound (95% lower confidence limit for the benchmark dose); BMDS = benchmark dose software; TiO$_2$ = titanium dioxide.

*Response modeled: lung tumors excluding cystic keratinizing squamous lesions—from two studies of fine TiO$_2$ [Lee et al. 1985; Muhle et al. 1991] and one study of ultrafine TiO$_2$ [Heinrich et al. 1995].
†Acceptable model fit determined by $P > 0.05$.
‡BMDS did not converge. Model fitting and BMD and BMDL estimation were carried out as described in Wheeler and Bailer [2007].
**Average model, as described by Wheeler and Bailer [2007], based on the multistage, Weibull, and log-probit models. P-values are not defined in model averaging because the degrees of freedom are unknown.

Table 4–6. Estimated mean airborne mass concentrations of fine and ultrafine TiO$_2$ in humans and related human lung burdens (TiO$_2$* surface area dose) associated with 1/1000 excess risk of lung cancer after a 45-year working lifetime

Particle size and model fit to rat dose-response data for lung tumors§	Critical dose in human lungs†				Mean airborne exposure‡	
	Particle surface area (m²/lung)		Particle mass (g/lung)		MPPD (ICRP) lung model (mg/m³)	
	MLE	95% LCL	MLE	95% LCL	MLE	95% LCL
Fine TiO$_2$ (2.1 μm, 2.2 GSD; 6.68 m²/g):						
Gamma	98.9	16.5	14.8	2.5	14.0	2.3
Logistic	13.4	10.3	2.0	1.5	1.9	1.5
Log-logistic	85.5	10.0	12.8	1.5	12.1	1.4
Log-probit	109.6	19.5	16.4	2.9	15.5	2.8
Multistage	83.5	6.1	12.5	0.9	11.8	0.9
Probit	11.5	8.4	1.7	1.3	1.6	1.2
Quantal-linear	2.9	2.2	0.4	0.3	0.4	0.3
Quantal-quadratic	35.3	31.0	5.3	4.6	5.0	4.4
Weibull	81.2	13.8	12.2	2.1	11.5	1.9
MA¶	93.5	17.0	14.0	2.5	13.2	2.4
*Ultrafine TiO$_2$ (0.8 μm, 1.8 GSD; 48 m²/g)***						
Gamma	98.9	16.5	2.06	0.34	1.71	0.28
Logistic	13.4	10.3	0.28	0.22	0.23	0.18
Log-logistic	85.5	10.0	1.78	0.21	1.48	0.17
Log-probit	109.6	19.5	2.28	0.41	1.89	0.34
Multistage	83.5	6.1	1.74	0.13	1.44	0.11
Probit	11.5	8.4	0.24	0.18	0.20	0.15
Quantal-linear	2.9	2.2	0.06	0.05	0.05	0.04
Quantal-quadratic	35.3	31.0	0.73	0.65	0.61	0.54
Weibull	81.2	13.8	1.69	0.29	1.40	0.24
MA¶	93.5	17.0	1.95	0.35	1.62	0.29

*Abbreviations: MA = model average; BMD = benchmark dose; GSD = geometric standard deviation, LCL = lower confidence limit; MLE = maximum likelihood estimate; TiO$_2$ = titanium dioxide, MPPD = multiple-path particle dosimetry [CIIT and RIVM 2002] model.

†MLE and 95% LCL were determined in rats (Table 4–5) and extrapolated to humans based on species differences in lung surface area, as described in Section 4.2.3.

‡Mean concentration estimates were derived from the CIIT and RIVM [2002] lung model.

§Without keratinizing cystic lesions.

¶Model averaging combined estimates from the multistage, Weibull, and log-probit models [Wheeler and Bailer 2007].

**Mass median aerodynamic diameter (MMAD). Agglomerated particle size for ultrafine TiO$_2$ was used in the deposition model [CIIT and RIVM 2002]. Specific surface area was used to convert from particle surface area dose to mass dose; thus airborne particles with different specific surface areas would result in different mass concentration estimates from those shown here.

Table 4–7. Model average estimates[†] of mean airborne mass concentrations of fine and ultrafine TiO_2 in humans and related human lung burdens (TiO_2* surface area dose) associated with various levels of excess risk of lung cancer after a 45-year working lifetime

Particle size and lifetime added risk estimated from rat dose-response data for lung tumors[¶]	Critical dose in human lungs[‡]				Mean airborne exposure[§]	
	Particle surface area (m²/lung)		Particle mass (g/lung)		MPPD (ICRP) lung model (mg/m³)	
	MLE	95% LCL	MLE	95% LCL	MLE	95% LCL
Fine TiO_2 (2.1 µm, 2.2 GSD; 6.68 m²/g):						
1 in 500	114.2	24.9	17.1	3.7	16.1	3.5
1 in 1000	93.5	17.0	14.0	2.5	13.2	**2.4**[††]
1 in 2000	76.3	11.1	11.4	1.7	10.8	1.6
1 in 5000	57.5	6.2	8.6	0.9	8.1	0.9
1 in 10,000	46.4	3.8	6.9	0.6	6.5	0.5
1 in 100,000	21.4	0.5	3.2	0.1	3.0	0.1
Ultrafine TiO_2 (0.8 µm, 1.8 GSD; 48 m²/g)**:						
1 in 500	114.2	24.9	2.38	0.52	1.97	0.43
1 in 1000	93.5	17.0	1.95	0.35	1.62	**0.29**[††]
1 in 2000	76.3	11.1	1.59	0.23	1.32	0.19
1 in 5000	57.5	6.2	1.20	0.13	0.99	0.11
1 in 10,000	46.4	3.8	0.97	0.08	0.80	0.07
1 in 100,000	21.4	0.5	0.45	0.01	0.37	0.01

*Abbreviations: MA = model average; BMD = benchmark dose; GSD = geometric standard deviation, LCL = lower confidence limit; MLE = maximum likelihood estimate; TiO_2 = titanium dioxide, MPPD = multiple-path particle dosimetry [CIIT and RIVM 2002] model.

[†]Model averaging combined estimates from the multistage, Weibull, and log-probit models [Wheeler and Bailer 2007].

[‡]MLE and 95% LCL were determined in rats (Table 4–5) and extrapolated to humans based on species differences in lung surface area, as described in Section 4.2.3.

[§]Mean concentration estimates were derived from the CIIT and RIVM [2002] lung model.

[¶]Without keratinizing cystic lesions.

**Mass median aerodynamic diameter (MMAD). Agglomerated particle size for ultrafine TiO_2 was used in the deposition model [CIIT and RIVM 2002]. Specific surface area was used to convert from particle surface area dose to mass dose; thus airborne particles with different specific surface areas would result in different mass concentration estimates from those shown here.

[††]The exposure levels shown in boldface are the 95% LCL estimates of the concentrations of fine and ultrafine TiO_2 considered appropriate for establishment of a REL. The ultrafine exposure level of 0.29 mg/m³ was rounded to 0.3 for the REL.

particle deposition of CIIT and RIVM [2002], which was used for particle dosimetry modeling in this analysis, would be an interstitialization/sequestration model that was developed and calibrated using data of U.S. coal miners [Kuempel et al. 2001a,b] and later validated using data of U.K. coal miners [Tran and Buchanan 2000]. The MPPD model uses the ICRP alveolar clearance model, which was developed using data on the clearance of radiolabeled tracer particles in humans, and has been in use for many years [ICRP 1994]. More data are needed to evaluate the model structures and determine how well each model would describe the retained doses associated with low particle exposures in humans, particularly for ultrafine particles. The MPPD model was selected for use in this analysis on the grounds that it is widely used and well accepted for particle dosimetry modeling in general, while use of the interstitialization/sequestration model has to date been limited primarily to modeling exposures to coal dust. Nevertheless, it should be noted that the interstitialization/sequestration model predicts lung burdens of fine and ultrafine TiO_2 which are approximately double those predicted by the MPPD model. Thus, the use of the interstitialization/sequestration model for particle dosimetry would approximately halve the estimates of occupational exposure levels equivalent to the rat critical dose levels, compared to estimates developed using the MPPD model.

The method selected for extrapolating between rats and humans also influences the estimates of occupational exposure levels equivalent to the rat critical dose levels. To extrapolate the critical particle surface area dose in the lungs of rats to the lungs of humans, either the relative mass or surface area of the lungs in each species could be used. The results presented in this analysis are based on the relative alveolar surface area, assuming 0.41 m^2 for Fischer 344 rats, 0.4 m^2 for Sprague-Dawley rats, and 102.2 m^2 for humans [Mercer et al. 1994]. Alternatively, extrapolation could be based on the relative lung weights of rat and human lungs, using a dose metric of particle surface area per gram of lung tissue. In that case, the estimates of the working lifetime occupational exposure levels equivalent to the rat critical dose levels would be higher by a factor of approximately four. The lung surface area-based approach was selected for this analysis because insoluble particles deposit and clear along the surface of the respiratory tract, so that dose per unit surface area is often used as a normalizing factor for comparing particle doses across species [EPA 1994].

The critical dose estimates in Table 4–5 also vary depending on the dose-response model used and on whether the MLE or the 95% LCL is used as the basis for estimation. The MLE dose estimates associated with a 1/1000 excess risk of lung tumors vary by a factor of 38, while the 95% LCL estimates vary by a factor of 14. All of the models provided statistically adequate fits to the data, and there is little basis to select one model over another. This uncertainty regarding model form has been addressed by the use of MA, as described by Wheeler and Bailer [2007], which weights the models based on the model fit as assessed by the Akaike information criterion. However, any of the models shown could conceivably be selected as the basis for human risk estimation. Use of the model-averaged 95% LCL value, as opposed to the model-averaged MLE, is intended to address both model uncertainty and variability in the rat data; however, it is possible that the 95% LCL may underestimate the true variability of the human population.

4.3.4 Mechanistic Considerations

The mechanism of action of TiO_2 is relevant to a consideration of the associated risks because, as discussed earlier, the weight of evidence suggests that the tumor response observed in rats exposed to fine and ultrafine TiO$_2$ results from a secondary genotoxic mechanism involving chronic inflammation and cell proliferation, rather than via genotoxicity of TiO$_2$ itself. This effect appears related to the physical form of the inhaled particle (i.e., particle surface area) rather than the chemical compound itself. Other PSLT particles—such as BaSO$_4$, carbon black, toner, and coal dust—also produce inflammation and lung tumors in proportion to particle surface area (Figures 3–2 and 3–3) and therefore appear to act via a similar mechanism.

Studies supporting this mechanism include empirical studies of the pulmonary inflammatory response of rats exposed to TiO$_2$ and other PSLT (see Sections 3.2.4 and 3.4.1); the tumor response of TiO$_2$ and other PSLT, which have consistent dose-response relationships (see Section 3.4.2); and in vitro studies, which show that inflammatory cells isolated from BALF from rats exposed to TiO$_2$ released ROS that could induce mutations in naive cells (see Section 3.2.1.3). There is some evidence, though limited, that inflammation may be a factor in the initiation of human lung cancer as well (see Section 3.5.2.2).

In considering all the data, NIOSH has determined that a plausible mechanism of action for TiO$_2$ in rats can be described as the accumulation of TiO$_2$ in the lungs, overloading of lung clearance mechanisms, followed by increased pulmonary inflammation and oxidative stress, cellular proliferation, and, at higher doses, tumorigenesis. These effects are better described by particle surface area than mass dose (see Section 3.4.2). The best-fitting dose-response curves for the tumorigenicity of TiO$_2$ are nonlinear; e.g., the multistage model in Table 4–5 is cubic with no linear term, the quantal-quadratic model is quadratic with no linear term, and the gamma and Weibull models have power terms of approximately 4 and 3, respectively. This nonlinearity is consistent with a secondary genotoxic mechanism and suggests that the carcinogenic potency of TiO$_2$ would decrease more than proportionately with decreasing *surface area* dose as described in the best-fitting risk assessment models.

4.4 Quantitative Comparison of Risk Estimates From Human and Animal Data

A quantitative comparison was performed (Appendix C) of the rat-based excess risk estimates for human lung cancer due to exposure to fine TiO$_2$ to the 95% upper confidence limit (UCL) of excess risk from the epidemiologic studies (Appendix D), in order to compare the rat- and human-based excess risks of lung cancer. If the sensitivity of the rat response to inhaled particulates differs from that of humans, then the excess risks derived from the rat data would be expected to differ from the excess risks estimated from the human studies. The results of the comparison of the rat- and human-based excess risk estimates were used to assess whether or not there was adequate precision in the data to reasonably exclude the rat model as a basis for predicting the excess risk of lung cancer in humans exposed to TiO$_2$.

The results of these comparisons showed that the MLE and 95% UCL excess risk estimates from the rat studies were lower than the 95% UCL from the human studies for an estimated working lifetime (Appendix C, Table C–1). These results indicate that, given the variability in the human studies [Fryzek et al. 2003; Boffetta et al. 2004], the rat-based excess risk estimates cannot reasonably be dismissed from use in predicting the excess risk of lung cancer in humans exposed to TiO$_2$. Thus, NIOSH determined that it is prudent to use these rat dose-response data for risk assessment in workers exposed to TiO$_2$.

4.5 Possible Bases for an REL

4.5.1 Pulmonary Inflammation

As discussed above, the evidence in rats suggests that the lung tumor mechanism associated with PSLT particles such as TiO$_2$ is a secondary genotoxic mechanism involving chronic inflammation and cell proliferation. One plausible approach to developing risk estimates for TiO$_2$ is to estimate exposure concentrations that would not be expected to produce an inflammatory response, thus preventing the development of responses that are secondary to inflammation, including cancer. A benchmark dose analysis for pulmonary inflammation in the rat was described in Section 4.3.1, and the results of extrapolating the rat BMDs to humans are presented in Table 4–3. Since two of the three studies available yielded 95% BMDLs of 0.78 and 1.03 mg/m³, a concentration of approximately 0.9 mg/m³ is reasonable as the starting point for development of recommendations for human exposures for fine TiO$_2$. Similarly, a concentration of approximately 0.11 mg/m³ is appropriate as the starting point for developing recommended exposures to ultrafine TiO$_2$.

As noted in Section 4.3.1.3, the human pulmonary inflammation BMDs in Table 4–3 are estimates of frank-effect levels and should be adjusted by the application of uncertainty factors to allow for uncertainty in animal-to-human extrapolation and interindividual variability. These uncertainty factors are commonly assumed to be ten-fold for animal-to-human extrapolation and another ten-fold for interindividual variability; the animal-to-human uncertainty may be subdivided into a factor of 4 for toxicokinetics and 2.5 for toxicodynamics (WHO 1994). Since the rat BMDs were extrapolated to humans using a deposition/clearance model, it is reasonable to assume that the animal-to-human toxicokinetic subfactor of 4 has already been accounted for; therefore, a total uncertainty factor of 25 (2.5 for animal-to-human toxicodynamics times 10 for interindividual variability) should be applied. This results in estimated exposure concentrations designed to prevent pulmonary inflammation of 0.04 mg/m³ for fine TiO$_2$ and 0.004 mg/m³ for ultrafine TiO$_2$.

4.5.2 Lung Tumors

Rather than estimating an exposure concentration designed to avoid secondary toxicity by preventing pulmonary inflammation, another possible basis for developing a REL is to model the risk of lung tumors directly. In the absence of mechanistic data in humans, the tumorigenic mechanism operative in rats cannot be ruled out. Therefore, one approach for estimation of recommended levels of occupational exposure to TiO$_2$ is to estimate the pulmonary particle surface area dose associated with a 1/1000 increase in rat lung tumors and to extrapolate that dose to humans on the basis of particle surface area per unit of lung surface area. This approach was used to assess the excess risk of lung cancer at various working lifetime exposure concentrations of fine or ultrafine TiO$_2$ (Table 4–6). Selection of the model for estimating risks has a significant impact on the risk estimates. As shown in Table 4–6, the 95% LCL working lifetime mean concentration of fine TiO$_2$ associated with a 1/1000 excess risk of lung cancer is 0.3 to 4.4 mg/m³, depending on the model used to fit the rat lung tumor data. For ultrafine TiO_2, the 95% LCL working lifetime mean concentration associated with a 1/1000 excess risk of lung cancer is 0.04 to 0.54 mg/m³, depending on the model.

Although any of the models evaluated in Table 4–6 could conceivably be used to develop recommendations for occupational exposures to TiO_2, the model averaging procedure is attractive since it incorporates both statistical variability and model uncertainty into confidence limit estimation. However, an argument could also be made for basing recommendations on the multistage model, due to its long history of use for carcinogen risk assessment or the quantal-linear model, on the grounds that it generates the lowest BMD and BMDL and is thus arguably the most health-protective. The BMD and BMDL derived via each of these models are shown in Table 4–6 for both fine and ultrafine TiO_2.

Since the various models produce different risk estimates and there is no clear mechanistically based preference for one model over another, it is appropriate to summarize the results by using a MA technique. MA, as implemented here, uses all the information from the various dose-response models, weighting each model by the Akaike information criterion for model fit and constructing an average dose-response model with lower bounds computed by bootstrapping. This method was described by Wheeler and Bailer [2007], who demonstrated via simulation studies that the MA method has superior statistical properties to a strategy of simply picking the best-fitting model from the BMDS suite. As shown in Table 4–6, the model average estimate of the working lifetime mean concentration of fine TiO_2 associated with a 1/1000 excess risk of lung cancer is 13.2 mg/m³, with a 95% LCL of 2.4 mg/m³. The corresponding estimates for ultrafine TiO_2 are 1.6 mg/m³, with a 95% LCL of 0.3 mg/m³. NIOSH believes that it is reasonable and prudent to use the 95% LCL model-averaged estimates as the basis for RELs, as opposed to the MLEs, in order to allow for model uncertainty and statistical variability in the estimates.

4.5.3 Comparison of Possible Bases for an REL

As discussed above, occupational exposure concentrations designed to prevent pulmonary inflammation, and thus prevent the development of secondary toxicity (including lung tumors), are 0.04 mg/m³ for fine TiO_2 and 0.004 mg/m³ for ultrafine TiO_2. In comparison, modeling of the dose-response relationship for lung tumors indicates that occupational exposure concentrations of 2.4 mg/m³ for fine TiO_2 and 0.3 mg/m³ for ultrafine TiO_2 would be sufficient to reduce the risk of lung tumors to a 1/1000 lifetime excess risk level. The discrepancy between the occupational exposure concentrations estimated from modeling either pulmonary inflammation or lung tumors raises serious questions concerning the optimal basis for a TiO_2 REL. However, it must be acknowledged that the two sets of possible RELs are not based on entirely comparable endpoints. The pulmonary inflammation-based exposure concentrations are expected to entirely prevent the development of toxicity secondary to pulmonary inflammation, resulting in zero excess risk of lung tumors due to exposure to TiO_2. In contrast, the lung tumor-based exposure concentrations are designed to allow a small, but nonzero, excess risk of lung tumors due to occupational exposure to TiO_2.

As discussed in Section 3.4.1, particle-induced pulmonary inflammation may act as a precursor for lung tumor development; however, pulmonary inflammation itself is not a specific biomarker for lung cancer. As noted in Section 3.5.2.2, the precise level of sustained inflammation necessary to initiate a tumorigenic response is currently unknown. It is possible that the 4% PMN response used in this analysis as the benchmark response level for pulmonary inflammation is overly protective and that a somewhat greater inflammatory response is required for tumor initiation.

It is also possible that the 25-fold uncertainty factor applied to the critical dose estimate for pulmonary inflammation may be overly conservative, since pulmonary inflammation is an early event in the sequence of events leading to lung tumors. However, NIOSH has not previously used early events or secondary toxicity as a rationale for applying smaller than normal uncertainty factors. Given that in this case the primary objective of preventing pulmonary inflammation is to prevent the development of lung tumors, and given that lung tumors can be adequately controlled by exposures many-fold higher than the inflammation-based exposure concentrations, NIOSH has concluded that it is appropriate to base RELs for TiO_2 on lung tumors rather than pulmonary inflammation. However, NIOSH notes that extremely low-level exposures to TiO_2—i.e., at concentrations less than the pulmonary inflammation-based RELs—may pose no excess risk of lung tumors.

5 Hazard Classification and Recommended Exposure Limits

NIOSH initiated the evaluation of titanium dioxide by considering it as a single substance with no distinction regarding particle size. However, a review of all the relevant scientific literature indicated that there could be a greater occupational health risk with smaller size (ultrafine) particles and therefore NIOSH provides separate recommendations for the ultrafine and fine categories.

NIOSH has reviewed the relevant animal and human data to assess the carcinogenicity of titanium dioxide (TiO_2) and has reached the following conclusions. First, the weight of evidence suggests that the tumor response observed in rats exposed to ultrafine TiO_2 resulted from a secondary genotoxic mechanism involving chronic inflammation and cell proliferation, rather than via direct genotoxicity of TiO_2. This effect appears to be related to the physical form of the inhaled particle (i.e., particle surface area) rather than to the chemical compound itself. Second, based on the weight of the scientific data (including increase in adenocarcinoma tumor incidence in a chronic inhalation study in rats of 10 mg/m³), NIOSH determined that inhaled ultrafine TiO_2 is a potential occupational carcinogen and is recommending exposure limits to minimize the cancer risk from exposure to ultrafine TiO_2. Finally, because the tumorigenic dose of fine TiO_2 (250 mg/m³) in the Lee et al. studies [1985, 1986a] was substantially higher than current inhalation toxicology practice—and because there was no significant increase in tumors at 10 or 50 mg/m³—NIOSH did not use the highest dose in its hazard identification and concluded that there is insufficient evidence to classify fine TiO_2 as a potential occupational carcinogen. Although NIOSH has determined that the data are insufficient for cancer hazard classification of fine TiO_2, the particle surface area dose and tumor response relationship is consistent with that observed for ultrafine TiO_2 and warrants that precautionary measures be taken to protect the health of workers exposed to fine TiO_2. Therefore, NIOSH used all of the animal tumor response data to conduct the dose-response modeling, and developed separate mass-based RELs for ultrafine and fine TiO_2.

5.1 Hazard Classification

NIOSH reviewed the current scientific data on TiO_2 to evaluate the weight of the evidence for the NIOSH designation of TiO_2 as a "potential occupational carcinogen." Two factors were considered in this evaluation: (1) the evidence in humans and animals for an increased risk of lung cancer from inhalation of TiO_2, including exposure up to a full working lifetime, and (2) the evidence on the biologic mechanism of the dose-response relationship observed in rats, including evaluation of the particle characteristics and dose metrics that are related to the pulmonary effects.

No exposure-related increase in carcinogenicity was observed in the epidemiologic studies conducted on workers exposed to TiO_2 dust in the workplace [Boffetta et al. 2001, 2003, 2004; Fryzek 2004; Fryzek et al. 2003, 2004a]. In rats exposed to fine TiO_2 by chronic inhalation, lung tumors were elevated at 250 mg/m³, but not at 10 or 50 mg/m³ [Lee et al. 1985; 1986a]. In contrast, chronic inhalation exposures to ultrafine TiO_2 at approximately 10 mg/m³ resulted in a statistically significant increase in malignant lung tumors in rats, although lung tumors in mice were not elevated [Heinrich et al. 1995]. The lung tumors observed in rats after exposure to 250 mg/m³ of fine TiO_2 were the basis for the original NIOSH designation of TiO_2 as a "potential occupational carcinogen." However, because this dose is considered to be significantly higher than currently accepted inhalation toxicology practice [Lewis et al. 1989], NIOSH concluded that the response at such a high dose should not be used in making its hazard identification. Therefore, NIOSH has come to different conclusions regarding the potential occupational carcinogenicity for fine versus ultrafine TiO_2. NIOSH evaluated the dose-response data in humans and animals, along with the mechanistic factors described below, in assessing the scientific basis for the current NIOSH designation of ultrafine but not fine TiO_2 as a "potential occupational carcinogen." In addition, NIOSH used the rat dose-response data in a quantitative risk assessment to develop estimates of excess risk of nonmalignant and malignant lung responses in workers over a 45-year working lifetime. These risk estimates were used in the development of RELs for fine and ultrafine TiO_2.

5.1.1 Mechanistic Considerations

As described in detail in Chapter 3, the mechanistic data considered by NIOSH were obtained from published subchronic and chronic studies in rodents exposed by inhalation to TiO_2 or other PSLT particles. These studies include findings on the kinetics of particle clearance from the lungs and on the nature of the relationship between particle surface area and pulmonary inflammation or lung tumor response. The mechanistic issues considered by NIOSH include the influence of particle size or surface area (vs. specific chemical reactivity) on the carcinogenicity of TiO_2 in rat lungs, the relationship between particle surface area dose and pulmonary inflammation or lung tumor response in rats, and the mechanistic evidence on the development of particle-elicited lung tumors in rats. These considerations are discussed in detail in Chapter 3.

NIOSH also considered the crystal structure as a modifying factor in TiO_2 carcinogenicity and inflammation. As described in Chapter 3, some short-term studies indicate that the particle surface properties pertaining to the crystal structure of TiO_2, including photoactivation, can influence the ROS generation, cytotoxicity, and acute lung responses. These studies also show that crystal structure does not influence the pulmonary inflammation or tumor responses following subchronic or chronic exposures [Bermudez et al. 2002, 2004]. The reason for these differences in the acute and longer-term responses with respect to TiO_2 crystal structure are not known but could relate to immediate effects of exposure to photoactivated TiO_2 and to quenching of ROS on the TiO_2 surfaces by lung surfactant. PSLT particles, including TiO_2, have relatively low surface reactivity compared to the more inherently toxic particles with higher surface activity such as crystalline silica [Duffin et al. 2007]. These findings are based on the studies in the scientific literature and may not apply to other formulations, surface coatings, or treatments of TiO_2 for which data were not available.

After analysis of the issues, NIOSH concluded that the most plausible mechanism for TiO_2 carcinogenesis is a nonchemical specific interaction of the particle with the cells in the lung, characterized by persistent inflammation and mediated by secondary genotoxic processes. The dose-response relationships for both the inflammation and tumorigenicity associated with TiO_2 exposure are consistent with those for other PSLT particles. Based on this evidence, NIOSH concluded that the adverse effects produced by TiO_2 exposure in the lungs are likely not substance-specific, but may be due to a nonchemical-specific effect of PSLT particles in the lungs at sufficiently high particle surface area exposures. However, because the tumorigenic dose for fine TiO_2 [Lee et al. 1985] of 250 mg/m³ was significantly higher than currently accepted inhalation toxicology practice [Lewis et al. 1989], NIOSH did not use the 250 mg/m³ dose in its hazard identification. Therefore, NIOSH concluded that there are insufficient data to classify fine TiO_2 as a potential occupational carcinogen but there are sufficient data indicating that ultrafine TiO_2 has the potential to cause cancer after adequate occupational exposure.

5.1.2 Limitations of the Rat Tumor Data

NIOSH evaluated all publicly available epidemiology studies and laboratory animal inhalation studies and determined that the best data to support a quantitative risk assessment for TiO_2 were from rat inhalation studies [Lee et al. 1985; Muhle et al. 1991; Heinrich et al. 1995; Cullen et al. 2002; Tran et al. 1999; Bermudez et al. 2002, 2004]. These studies provided exposure-response data for both inflammation and tumorigenicity and were used as the basis of the quantitative risk assessment.

NIOSH considered the scientific literature that supported and disputed on the rat as an appropriate model for human lung cancer after exposure to inhaled particles and reached the conclusion that the rat is an appropriate species on which to base its quantitative risk assessment for TiO_2. Although there is not extensive evidence that the overloading of lung clearance, as observed in rats (Chapter 3), occurs in humans, lung burdens consistent with overloading doses in rats have been observed in some humans with dusty jobs (e.g., coal miners) [Stöber et al. 1965; Carlberg et al. 1971; Douglas et al. 1986]. Rather than excluding the rat as the appropriate model, the lung overload process may cause the rat to attain lung burdens comparable to those that can occur in workers with dusty jobs. In addition, evidence suggests that, as in the rat, inhalation of particles increases the human inflammatory response, and increases in the inflammatory response may increase the risk of cancer (see Section 3.5.2.2). This information provides additional support for the determination that the rat is a reasonable animal model with which to predict human tumor response for particles such as TiO_2.

After evaluating all of the available data, NIOSH concluded that the appropriate dose metric in its risk assessment was particle surface area. Both tumorigenicity and inflammation were more strongly associated with particle surface area than particle mass dose. The separate risk estimates for fine and ultrafine TiO_2 are supported by the higher lung cancer potency in rats of ultrafine TiO_2 compared to fine TiO_2, which was associated with the greater surface area of ultrafine particles for a given mass. In rats chronically exposed to airborne fine TiO_2, statistically significant excess lung tumors were observed only in the 250 mg/m³ dose group. Although exposure concentrations greater than

100 mg/m³ are not currently standard methodology in inhalation toxicity studies [Lewis et al. 1989], and NIOSH questions the relevance of the 250 mg/m³ dose for classifying exposure to TiO_2 as a carcinogenic hazard to workers, the tumor-response data are consistent with that observed for ultrafine TiO_2 when converted to a particle surface area metric. Thus to be cautious, NIOSH used all of the animal tumor response data when conducting dose-response modeling and determining separate RELs for ultrafine and fine TiO_2. With chronic exposure to airborne ultrafine TiO_2, lung tumors were seen in rats exposed to an average of approximately 10 mg/m³ (ranged from 7.2 mg/m³ to 14.8 mg/m³) over the exposure period.

It would be a better reflection of the entire body of available data to set RELs as the inhaled surface area of the particles rather than the mass of the particles. This would be consistent with the scientific evidence showing an increase in potency with increase in particle surface area (or decrease in particle size) of TiO_2 and other PSLT particles. For this reason, the basis of the RELs for fine and ultrafine TiO_2 is the rat dose-response data for particle surface area dose and pulmonary response. However, current technology does not permit the routine measurement of the surface area of airborne particles; therefore, NIOSH recommends sampling the mass airborne concentration of TiO_2 as two broad primary particle size categories: fine (< 10 μm) and ultrafine (< 0.1 μm). These categories reflect current aerosol size conventions, although it is recognized that actual particle size distributions in the workplace will vary.

5.1.3 Cancer Classification in Humans

Since the public comment and peer review draft of this document was made available, NIOSH has learned that the IARC has reassessed TiO_2. IARC now classifies TiO_2 as an IARC Group 2B carcinogen, "possibly carcinogenic to humans" [IARC 2010]. NIOSH supports this decision and the underlying analysis leading to this conclusion.

Based on the animal studies described in Chapter 3 and the quantitative risk assessment in Chapter 4, NIOSH has concluded that ultrafine but not fine TiO_2 particulate matter is a potential occupational carcinogen. However, as a precautionary step, NIOSH conducted a quantitative risk assessment based on the tumor data for both fine and ultrafine TiO_2.

The potency of ultrafine TiO_2, which has a much higher surface area per unit mass than fine TiO_2, was many times greater than fine TiO_2, with malignant tumors observed at the lowest dose level of ultrafine TiO_2 tested (10 mg/m³).

The lack of an exposure-response relationship in the epidemiologic studies of workers exposed to TiO_2 dust in the workplace should not be interpreted as evidence of discordance between the mechanism presumed to operate in rats and the human potential for carcinogenicity. As demonstrated by the quantitative comparison between the animal and human studies (see Section 4.4 and Appendix C), the responses were not statistically inconsistent: the epidemiologic studies were not sufficiently precise to determine whether they replicated or refuted the animal dose-response. This is not a surprising finding for carcinogens of lower potency, such as fine-sized TiO_2.

The mechanistic data reviewed above, however, leave open the possibility of species differences beyond what would be anticipated for a genotoxic carcinogen. Although it is plausible that the secondary genotoxic mechanism proposed here operates in humans exposed to TiO_2 dust, there is insufficient evidence to corroborate this. In addition, there is limited information on the kinetics or specific physiological response to TiO_2 particles in humans. The evidence suggests that inhalation of lower surface area TiO_2 is not likely to result in carcinogenicity in any test species. This concept is reflected in the quantitative risk assessment, in which the curvilinear dose response predicts that lower exposures have disproportionally less risk than higher exposures. For workers, this suggests that exposure to concentrations lower than the RELs will be less hazardous and may pose a negligible risk.

Although the analysis in this document is limited to consideration of the workplace hazard posed by TiO_2, these findings suggest that other PSLT particles inhaled in the workplace may pose similar hazards, particularly nano-sized particles with high surface areas. NIOSH is concerned that other nano-sized PSLT particles may have similar health effects to those observed for TiO_2, in which ultrafine TiO_2 particles were observed to be more carcinogenic and inflammogenic on a mass basis than fine TiO_2 [Heinrich et al. 1995; Lee et al. 1985, 1986a].

5.2 Recommended Exposure Limits

NIOSH recommends airborne exposure limits of 2.4 mg/m³ for fine TiO_2 and 0.3 mg/m³ for ultrafine (including engineered nanoscale) TiO_2 as TWA concentrations for up to 10 hr/day during a 40-hour work week, using the international definitions of respirable dust [CEN 1993; ISO 1995] and the NIOSH Method 0600 for sampling airborne respirable particles [NIOSH 1998]. NIOSH selected these exposure limits for recommendation because they would reduce working lifetime risks for lung cancer to below 1/1000. Cancer risks greater than 1/1000 are considered significant and worthy of intervention by OSHA. NIOSH has used this risk level in a variety of circumstances, including citing this level as appropriate for developing authoritative recommendations in Criteria Documents and risk assessments published in scientific journal articles [NIOSH 1995; NIOSH 2006; Rice et al. 2001; Park et al. 2002; Stayner et al. 2000; Dankovic et al. 2007]. It is noted that the true risk of lung cancer due to exposure to TiO_2 at these concentrations could be much lower than 1/1000. To account for the risk that exists in work environments where airborne exposures to fine and ultrafine TiO_2 occur, exposure measurements to each size fraction should be combined using the additive formula and compared to the additive REL of 1 (unitless) (see Figure 6–1 Exposure assessment protocol for TiO_2).

Because agglomerated ultrafine particles are frequently measured as fine-sized but behave biologically as ultrafine particles due to the surface area of the constituent particles, exposures to agglomerated ultrafine particles should be controlled to the ultrafine REL.

"Respirable" is defined as particles of aerodynamic size that, when inhaled, are capable of depositing in the gas-exchange (alveolar) region of the lungs [ICRP 1994]. Sampling methods have been developed to estimate the airborne mass concentration of respirable particles [CEN 1993; ISO 1995; NIOSH 1998]. "Fine" is defined in this document as all particle sizes that are collected by respirable particle sampling (i.e., 50% collection efficiency for particles of 4 µm, with some collection of particles up to 10 µm). "Ultrafine" is defined as the fraction of respirable particles with primary particle diameter < 0.1 µm (< 100 nm), which is a widely used definition. Additional methods are needed to determine if an airborne respirable particle sample includes ultrafine TiO_2 (Chapter 6).

The NIOSH RELs for fine TiO_2 of 2.4 mg/m³ and ultrafine TiO_2 of 0.3 mg/m³ were derived using the model averaging procedure (described in Sections 4.3–4.5) of the dose-response relationship for lung tumors in rats, extrapolated to occupational exposures. The ultrafine REL reflects NIOSH's greater concern for the potential carcinogenicity of ultrafine TiO_2 particles. As particle size decreases, the surface area increases (for equal mass), and the tumor potency increases per mass unit of dose. NIOSH acknowledges the evolving state of knowledge on the hazards associated with nano-sized particles and invites additional discussion on the risks of nanoparticulate exposures in the workplace.

6 Measurement and Control of TiO_2 Aerosol in the Workplace

6.1 Exposure Metric

Based on the observed relationship between particle surface area dose and toxicity (Chapters 3 and 4), the measurement of aerosol surface area would be the preferred method for evaluating workplace exposures to TiO_2. However, personal sampling devices that can be routinely used in the workplace for measuring particle surface area are not currently available. As an alternative, if the airborne particle size distribution of the aerosol is known in the workplace and the size distribution remains relatively constant with time, mass concentration measurements may be useful as a surrogate for surface area measurements. NIOSH is recommending that a mass-based airborne concentration measurement be used for monitoring workplace exposures to fine and ultrafine (including engineered nanoscale) TiO_2 until more appropriate measurement techniques can be developed. NIOSH is currently evaluating the efficacy of various sampling techniques for measuring fine and ultrafine TiO_2 and may make specific recommendations at a later date.

In the interim, personal exposure concentrations to fine (pigment-grade) and ultrafine (including engineered nanoscale) TiO_2 should be determined with NIOSH Method 0600 using a standard 10-mm nylon cyclone or equivalent particle size-selective sampler [NIOSH 1998]. Measurement results from NIOSH Method 0600 should provide a reasonable estimate of the exposure concentration to fine and ultrafine (including engineered nanoscale) TiO_2 at the NIOSH RELs of 2.4 and 0.3 mg/m^3, respectively, when the predominant exposure to workers is TiO_2. No personal sampling devices are available at this time to specifically measure the mass concentrations of ultrafine aerosols; however, the use of NIOSH Method 0600 will permit the collection of most airborne ultrafine (including engineered nanoscale) particles and agglomerates.

In work environments where exposure to other types of aerosols occur or when the size distribution of TiO_2 (fine vs. ultrafine) is unknown, other analytical techniques may be needed to characterize exposures. NIOSH Method 7300 [NIOSH 2003] can be used to assist in differentiating TiO_2 from other aerosols collected on the filter while electron microscopy, equipped with X-ray energy dispersive spectroscopy (EDS), may be needed to measure and identify particles. In workplaces where TiO_2 is purchased as a single type of bulk powder, the primary particle size of the bulk powder can be used to determine whether the REL for fine or ultrafine should be applied if adequate airborne exposure data exist to confirm that the airborne particle size has not substantially been altered during the handling and/or material processing of TiO_2. Although NIOSH Methods 0600 and 7300 have not been validated in the field for airborne exposure to TiO_2, they have been validated for other substances and, therefore, should provide results for TiO_2 within the expected accuracy of the Methods.

6.2 Exposure Assessment

A multitiered workplace exposure assessment might be warranted in work environments where the airborne particle size distribution of TiO_2 is unknown (fine vs. ultrafine) and/or where other airborne aerosols may interfere with the interpretation of sample results. Figure 6–1 illustrates an exposure assessment strategy that can be used to measure and identify particles so that exposure concentrations can be determined for fine and ultrafine (including engineered nanoscale) TiO_2. An initial exposure assessment should include the simultaneous collection of respirable dust samples with one sample using a hydrophobic filter (as described in NIOSH Method 0600) and the other sample using a mixed cellulose ester filter (MCEF).* If the respirable exposure concentration for TiO_2 (as determined by Method 0600) is less than 0.3 mg/m³, then no further action is required. If the exposure concentration exceeds 0.3 mg/m³, then additional characterization of the sample is needed to determine the particle size and percentage of TiO_2 and other extraneous material on the sample. To assist in this assessment, the duplicate respirable sample collected on a MCEF should be evaluated using transmission electron microscopy (TEM) to size particles and determine the percentage of fine (> 0.1 μm) and ultrafine (< 0.1 μm) TiO_2. When feasible, the percent of fine and ultrafine TiO_2 should be determined based on the measurement of "primary" particles (includes agglomerates comprised of primary particles). The identification of particles (e.g., TiO_2) by TEM can be accomplished using EDS or energy loss spectroscopy. Once the percent of TiO_2 (by particle size) has been determined, adjustments can be made to the mass concentration (as determined by Method 0600) to assess whether exposure to the NIOSH RELs for fine, ultrafine (including engineered nanoscale), or combined fine and ultrafine TiO_2 had been exceeded.

To minimize the future need to collect samples for TEM analysis, samples collected using Method 0600 can also be analyzed using NIOSH Method 7300 or other equivalent methods to determine the amount of TiO_2 on the sample. When using NIOSH Method 7300, it is important that steps be taken (i.e., pretreatment with sulfuric or hydrofluoric acid) to insure the complete dissolution and recovery of TiO_2 from the sample. The results obtained using Method 7300, or other equivalent method, should be compared with the respirable mass concentration to determine the relative percentage of TiO_2 in the concentration measurement. Future assessments of workplace exposures can then be limited to the collection and analysis of samples using Method 0600 and Method 7300 (or equivalent) to ensure that airborne TiO_2 concentrations have not changed over time.

6.3 Control of Workplace Exposures to TiO_2

Given the extensive commercial use of fine (pigment grade) TiO_2, the potential for occupational exposure exists in many workplaces. However, few data exist on airborne concentrations and information on workplaces where exposure might occur. Most of the available data for fine TiO_2 are reported as total dust and not as the respirable fraction. Historical total dust exposure measurements collected in TiO_2 production plants in the United States

---

*Note: The collection time for samples using a MCEF may need to be shorter than the duplicate samples collected and analyzed by Method 0600 to ensure that particle loading on the filter doesn't become excessive and hinder particle sizing and identification by TEM.

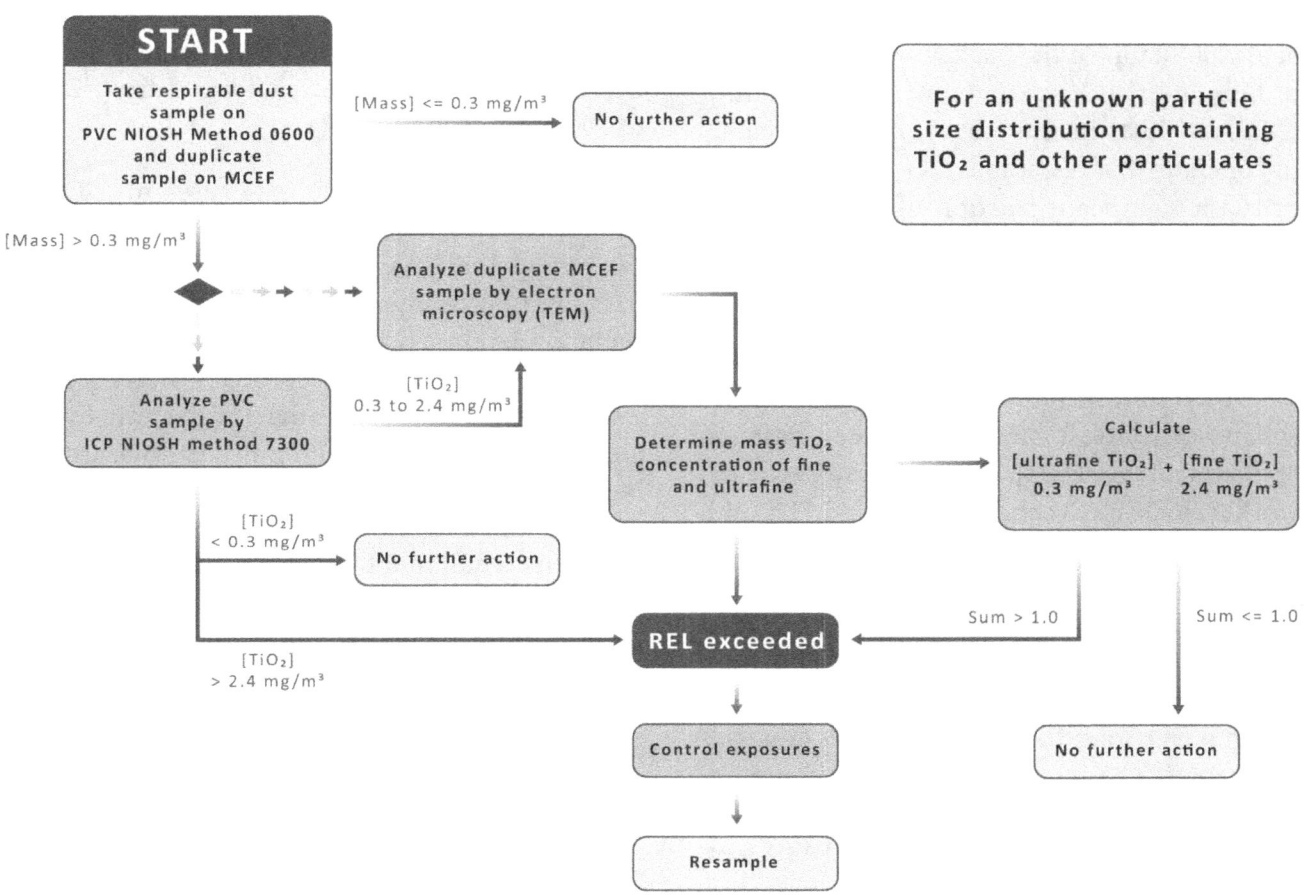

Figure 6–1. Exposure assessment protocol for TiO$_2$.

often exceeded 10 mg/m³ [IARC 1989], while more contemporary exposure concentrations in these plants indicate that mean total inhalable dust measurements may be below 3 mg/m³ (1.1 mg/m³ median) for most job tasks [Fryzek et al. 2003]. Given the particle size dimensions of fine TiO_2 (~0.1 μm to 4 μm, avg. of 0.5 μm) [Malvern Instruments 2004], it is reasonable to conclude that respirable TiO_2 particles comprised a significant fraction of the total dust measurement. Estimates of worker exposures to respirable TiO_2 determined in 1999 from 8 European plants producing fine TiO_2 ranged from 0.1 to 5.99 mg/m³ (plant with highest measured concentrations) to 0.1 to 0.67 mg/m³ (plant with lowest measured concentrations) [Boffetta et al. 2003]. The highest reported worker exposure concentrations to TiO_2 in both U.S. and European production plants were among packers and micronizers and during maintenance [Sleeuwenhoek 2005]. Results of a NIOSH health hazard evaluation conducted at a facility using powder paint containing TiO_2 found airborne concentrations of respirable TiO_2 ranging from 0.01 to 1.0 mg/m³ (9 samples) [NIOSH 2009b]. TEM analysis conducted on one airborne sample found 42.4% of the TiO_2 particles less than 0.1 μm in diameter. NIOSH is not aware of any extensive commercial production of ultrafine anatase TiO_2 in the United States although it may be imported for use. Ultrafine rutile TiO_2 is being commercially produced as an additive for plastics to absorb and scatter ultraviolet light; 10%–20% of the ultrafine TiO_2 is reported to be < 100 nm in size [DuPont 2007]. Engineered TiO_2 nanoparticles are also being manufactured, and like ultrafine TiO_2, they are finding commercial application as a photocatalyst for the destruction of chemical and microbial contaminants in air and water, in light-emitting diodes and solar cells, in plastics, as a UV blocker, and as a "self-cleaning" surface coating. While a paucity of data exist on worker exposure to engineered TiO_2, exposure measurements taken at a facility manufacturing engineered TiO_2 found respirable exposure concentrations as high as 0.14 mg/m³ [Berges et al. 2007]. The primary particle size determined by TEM analysis ranged from 25 to 100 nm.

The control of workplace exposures to TiO_2 should be primarily accomplished through the use of engineering controls. Although limited data exist on occupational exposures to TiO_2, reducing exposures can be achieved using a variety of standard control techniques [Raterman 1996; Burton 1997]. Standard industrial hygiene practices for controlling airborne hazards include engineering controls, work practice and administrative procedures, and personal protective equipment. Examples of engineering controls include process modifications and the use of an industrial ventilation system to reduce worker airborne exposures [ACGIH 2001c]. In general, control techniques such as source enclosure (i.e., isolating the generation source from the worker) and local exhaust ventilation systems are the preferred methods for preventing worker exposure to aerosols. In light of current scientific knowledge regarding the generation, transport, and capture of aerosols, these control techniques should be effective for both fine and ultrafine particles [Seinfeld and Pandis 1998; Hinds 1999]. Conventional engineering controls, using ventilation systems to isolate the exposure source from workers, should be effective in reducing airborne exposures to fine and ultrafine (including engineered nanoscale) TiO_2. Ventilation systems should be equipped with high efficiency particulate air (HEPA) filters which are designed to remove 99.97% of particles 300 nm in diameter. Particles smaller than 200 nm are generally collected on the filter by diffusion, whereas particles larger than

800 nm are deposited through impaction and interception [Lee and Liu 1981, 1982]. Ventilation systems must be properly designed, tested, and routinely maintained to provide maximum efficiency.

When engineering controls and work practices cannot reduce worker TiO_2 exposures to below the REL then a respirator program should be implemented. The OSHA respiratory protection standard (29 CFR 1910.134) sets out the elements for both voluntary and required respirator use. All elements of the standard should be followed. Primary elements of the OSHA respiratory protection standard include (1) an evaluation of the worker's ability to perform the work while wearing a respirator, (2) regular training of personnel, (3) periodic environmental monitoring, (4) respirator fit-testing, and (5) respirator maintenance, inspection, cleaning, and storage. The program should be evaluated regularly and respirators should be selected by the person who is in charge of the program and knowledgeable about the workplace and the limitations associated with each type of respirator.

NIOSH provides guidance for selecting an appropriate respirator in NIOSH Respirator Selection Logic 2004 [NIOSH 2004] available online at http://www.cdc.gov/niosh/docs/2005-100/default.html. The selection logic takes into account the expected exposure concentration, other potential exposures, and the job task. For most job tasks involving TiO_2 exposure, a properly fit-tested, half-facepiece particulate respirator will provide protection up to 10 times the respective REL. When selecting the appropriate filter and determining filter change schedules, the respirator program manager should consider that particle overloading of the filters may occur because of the size and characteristics of TiO_2 particles. Studies on the filtration performance of N95 filtering-facepiece respirators found that the mean penetration levels for 40 nm particles ranged from 1.4% to 5.2%, indicating that N95 and higher performing respirator filters would be effective at capturing airborne ultrafine TiO_2 particles [Balazy et al. 2006; Rengasamy et al. 2007, 2008].

In workplaces where there is potential worker exposure to TiO_2, employers should establish an occupational health surveillance program. Occupational health surveillance, which includes hazard surveillance (hazard and exposure assessment) and medical surveillance, is an essential component of an effective occupational safety and health program. An important objective of the program is the systematic collection and analysis of exposure and health data for the purpose of preventing illness. In workplaces where exposure to ultrafine or engineered TiO_2 occurs, NIOSH has published interim guidance on steps that can be taken to minimize the risk of exposure [NIOSH 2009a] and recommendations for medical surveillance and screening [NIOSH 2009c] that could be used in establishing an occupational health surveillance program.

7 Research Needs

Additional data and information are needed to assist NIOSH in evaluating the occupational safety and health issues of working with fine and ultrafine TiO_2. Data are particularly needed on the airborne particle size distributions and exposures to ultrafines in specific operations or tasks. These data may be merged with existing epidemiologic data to determine if exposure to ultrafine TiO_2 is associated with adverse health effects. Information is needed about whether respiratory health (e.g., lung function) is affected in workers exposed to TiO_2. Experimental studies on the mechanism of toxicity and tumorigenicity of ultrafine TiO_2 would increase understanding of whether factors in addition to surface area may be important. Although sampling devices for all particle sizes are available for research purposes, practical devices for routine sampling in the workplace are needed.

7.1 Workplace Exposures and Human Health

- Quantify the airborne particle size distribution of TiO_2 by job or process and obtain quantitative estimates of workers' exposures to fine and ultrafine TiO_2.

- Conduct epidemiologic studies of workers manufacturing or using TiO_2-containing products using quantitative estimates of exposure by particle size, including fine and ultrafine fractions (see bullet above).

- Evaluate the extent to which the specific surface area in bulk TiO_2 is representative of the specific surface area of the airborne TiO_2 particles that workers inhale and that are retained in the lungs.

- Investigate the adequacy of current mass-based human lung dosimetry models for predicting the clearance and retention of inhaled ultrafine particles.

7.2 Experimental Studies

- Investigate the fate of ultrafine particles (e.g., TiO_2) in the lungs and the associated pulmonary responses.

- Investigate the ability of ultrafine particles (e.g., TiO_2) to enter cells and interact with organelle structures and DNA in mitochondria or the nucleus.

7.3 Measurement, Controls, and Respirators

- Develop accurate, practical sampling devices for ultrafine particles (e.g., surface area sampling devices).

- Evaluate effectiveness of engineering controls for controlling exposures to fine and ultrafine TiO_2.

- Determine effectiveness of respirators for ultrafine TiO_2.

References

ACGIH [1984]. Particle size-selective sampling in the workplace. Report of the ACGIH Technical Committee on air sampling procedures. Ann Am Conf Gov Ind Hyg 11:23–100.

ACGIH [1994]. 1994–1995 Threshold limit values for chemical substances and physical agents and biological exposure indices. Cincinnati, OH: Americal Conference of Governmental Industrial Hygenists.

ACGIH [2001a]. Industrial ventilation: a manual of recommended practice. 24th ed. Cincinnati, OH: American Conference of Governmental Industrial Hygenists.

ACGIH [2001b]. Particulates (insoluble) not otherwise specified (PNOS). In: Documentation of the threshold limit values for chemical substances. 7th ed. Cincinnati, OH: American Conference of Governmental Industrial Hygienists.

ACGIH [2001c]. Titanium dioxide. In: Documentation of the threshold limit values for chemical substances. 7th ed. Cincinnati, OH: American Conference of Governmental Industrial Hygienists.

ACGIH [2005]. 2005 TLVs® and BEIs® based on the documentation of the threshold limit values for chemical substances and physical agents and biological exposure indices. Cincinnati, OH: American Conference of Governmental Industrial Hygenists.

ACGIH [2009]. 2009 TLVs® and BEIs® based on the documentation of the threshold limit values for chemical substances and physical agents and biological exposure indices. Cincinnati, OH: American Conference of Governmental Industrial Hygienists.

Adamson IY, Bowden DH [1981]. Dose response of the pulmonary macrophagic system to various particulates and its relationship to transepithelial passage of free particles. Exp Lung Res 2(3):165–175.

Adamson IY, Hedgecock C [1995]. Patterns of particle deposition and retention after instillation to mouse lung during acute injury and fibrotic repair. Exp Lung Res 21(5):695–709.

Aitken RJ, Creely KS, Tran CL [2004]. Nanoparticles: an occupational hygiene review. HSE Research Report 274. United Kingdom: Health & Safety Executive [http://www.hse.gov.uk/research/rrhtm/rr274.htm].

American Chemistry Council [2006]. Comments of the titanium dioxide panel of the American Chemistry Council and the titanium dioxide manufacturers association and the physical sunscreen manufacturers association of the European Chemical Industry Council on NIOSH current intelligence bulletin: evaluation of health hazard and recommendations for occupational exposure to titanium dioxide. March 31, 2006. p. 7 [http://www.cdc.gov/niosh/docket/pdfs/NIOSH-033/Submissions/0033-033106-ACC%20TIO2_submission.pdf].

Attfield MD, Costello J [2004]. Quantitative exposure-response for silica dust and lung cancer in Vermont granite workers. Am J Ind Med 45:129–138.

Baan RA [2007]. Carcinogenic hazards from inhaled carbon black, titanium dioxide, and talc not containing asbestos or asbestiform fibers: recent evaluations by an IARC Monographs Working Group. Inhal Toxicol 19(Suppl. 1):213–228.

Baggs RB, Ferin J, Oberdörster G [1997]. Regression of pulmonary lesions produced by inhaled titanium dioxide in rats. Vet Pathol 34:592–597.

Bailer AJ, Stayner LT, Smith RJ, Kuempel ED, Prince, MM [1997]. Estimating benchmark concentrations and other non-cancer endpoints in epidemiology studies. Risk Anal 17(6):771–780.

Bałazy A, Toivola M, Reponen T, Podgórski A, Zimmer A, Grinshpun SA [2006]. Manikin-based performance evaluation of N95 filtering-facepiece respirators challenged with nanoparticles. Ann Occup Hyg 50:259–269.

Beaumont JJ, Sandy MS, Sherman CD [2004]. Titanium dioxide and lung cancer [letter to the editor]. J Occup Environ Med 46(8):759.

Behnajady MA, Modirshahla N, Shokri M, Elham H, Zeininezhad A [2008]. The effect of particle size and crystal structure of titanium dioxide nanoparticles on the photocatalytic properties. J Environ Sci Health A Tox Hazard Subst Environ Eng 43(5):460–467.

BEIR IV [1998]. Health risks of radon and other internally deposited alpha-emitters (BEIR IV). Washington, DC: National Academy Press, pp. 131–136.

Bellmann B, Muhle H, Creutzenberg O, Dasenbrock C, Kilpper R, MacKenzie JC, Morrow P, Mermelstein R [1991]. Lung clearance and retention of toner, utilizing a tracer technique, during chronic inhalation exposure in rats. Fund Appl Toxicol 17:300–313.

Berges M, Möhlmann C, Swennen B, van Rompaey Y, Berghmans P [2007]. Workplace exposure characterization at TiO_2 nanoparticle production. In: Proceedings of the 3rd International Symposium on Nanotechnology, Occupational and Environmental Health, Taipei, Taiwan, August 29.

Bermudez E, Mangum JB, Asgharian B, Wong BA, Reverdy EE, Janszen DB, Hext PM, Warheit DB, Everitt JI [2002]. Long-term pulmonary responses of three laboratory rodent species to subchronic inhalation of pigmentary titanium dioxide particles. Toxicol Sci 70(1):86–97.

Bermudez E, Mangum JB, Wong BA, Asgharian B, Hext PM, Warheit DB, Everitt JI [2004]. Pulmonary responses of mice, rats, and hamsters to subchronic inhalation of ultrafine titanium dioxide particles. Toxicol Sci 77:347–357.

Bernard BK, Osheroff MR, Hofmann A, Mennear JH [1990]. Toxicology and carcinogenesis studies of dietary titanium dioxide-coated mica in male and female Fischer 344 rats. J Toxicol Environ Health 29(4):417–429.

BLS [2006]. May 2006 National industry-specific occupational employment and wage estimates. Washington, DC: U.S. Department of Labor, Bureau of Labor Statistics. [http://www.bls.gov/oes/2006/may/oessrci.htm#31-33].

Boffetta P, Gaborieau V, Nadon L, Parent M-E, Weiderpass E, Siemiatycki J [2001]. Exposure to titanium dioxide and risk of lung cancer in a population-based study from Montreal. Scand J Work Environ Health 27:227–232.

Boffetta P, Soutar A, Cherrie JW, Granath F, Andersen A, Anttila A, Blettner M, Gaborieau V, Klug SJ, Langard S, Luce D, Merletti F, Miller B, Mirabelli D, Pukkala E, Adami H-O, Weiderpass E [2004]. Mortality among workers employed in the titanium dioxide production industry in Europe. Cancer Causes Control 15(7):697–706.

Boffetta P, Soutar A, Weiderpass E, Cherrie J, Granath F, Andersen A, Anttila A, Blettner M, Gaborieau V, Klug S, Langard S, Luce D, Merletti F, Miller B, Mirabelli D, Pukkala E, Adami H-O [2003]. Historical cohort study of workers employed in the titanium dioxide production industry in Europe. Stockholm, Sweden: Karolinska Institute, Department of Medical Epidemiology. Unpublished.

Boorman GA, Brockman M, Carlton WW, Davis JMG, Dungworth DL, Hahn FF, Mohr U, Reichhelm H-BR, Turusov VS, Wagner BM [1996]. Classification of cystic keratinizing squamous lesions of the rat lung: report of a workshop. Toxicol Pathol 24:564–572.

Borm PJA, Höhr D, Steinfartz Y, Zeitträger I, Albrecht C [2000]. Chronic inflammation and tumor-formation in rats after intratracheal instillation of high doses of coal dusts, titanium dioxides, and quartz. Inhal Toxicol 12(Suppl 3):225–231.

BSI [2005]. Publicly available specification: Vocabulary—nanoparticles. London, UK: British Standards Institute. Document no. PAS 71:2005, pp. 32.

Burton DJ [1997]. General methods for the control of airborne hazards. In: DiNardi SR, ed. The occupational environment—its evaluation and control. Fairfax, VA: American Industrial Hygiene Association.

Carlberg JR, Crable JV, Limtiaca LP, Norris HB, Holtz JL, Mauer P, Wolowicz FR [1971]. Total dust, coal, free silica, and trace metal concentrations in bituminous coal miners' lungs. Am Ind Hyg Assoc J 32(7):432–440.

Carlton WW [1994]. Proliferative keratin cyst, a lesion in the lungs of rats following chronic exposure to para-aramid fibrils. Fundam Appl Toxicol 23(2):304–307.

Carter JM, Corson N, Driscoll KE, Elder A, Finkelstein JN, Harkema JN, Gelein R, Wade-Mercer P, Nguyen K, Oberdörster G [2006]. A comparative dose-related response of several key pro- and anti-inflammatory mediators in the lungs of rats, mice, and hamsters after subchronic inhalation of carbon black. J Occup Environ Med 48(12):1265–1278.

Castranova V [1998]. Particulates and the airways: basic biological mechanisms of pulmonary pathogenicity. Appl Occup Environ Hyg 13(8):613–616.

Castranova V [2000]. From coal mine dust to quartz: mechanisms of pulmonary pathogenicity. Inhal Toxicol 12 (Suppl 3):7–14.

CEN [1993].Workplace atmospheres—size fraction definitions for measurement of airborne particles, EN 481. Brussels, Belgium: European Committee for Standardization.

CFR. Code of Federal regulations. Washington, DC: U.S. Government Printing Office, Office of the Federal Register.

Chen JL, Fayerweather WE [1988]. Epidemiologic study of workers exposed to titanium dioxide. J Occup Med 30(12):937–942.

Chen H-W, Su S-F, Chien C-T, Lin W-H, Yu S-L, Chou C-C, Chen JJW, Yang P-C [2006]. Titanium dioxide nanoparticles induce emphysema-like lung injury in mice. FASEB J 20:E17312–E1741.

Cherrie JW [1999]. The effect of room size and general ventilation on the relationship between near and far-field concentrations. Appl Occup Environ Hyg 14(8):539–546.

Cherrie JW, Schneider T [1999]. Validation of a new method for structured subjective assessment of past concentrations. Ann Occup Hyg 43(4):235–246.

Cherrie JW, Schneider T, Spankie S, Quinn M [1996]. A new method for structured, subjective assessments of past concentrations. Occup Hyg 3:75–83.

CIIT and RIVM [2002]. Multiple-path particle dosimetry (MPPD V 1.0): a model for human and rat airway particle dosimetry. Research Triangle Park, NC: Chemical Industry Institute of Toxicology, Centers for Health Research. Bilthoven, The Netherlands: National Institute for Public Health and the Environment (RIVM) in the Netherlands.

Costabel U, Donner CF, Haslam PL, Rizzato G, Teschler H, Velluti G, Wallaert B [1990]. Clinical guidelines and indications for bronchoalveolar lavage (BAL): Occupational lung diseases due to inhalation of inorganic dust. Eur Respir J 3(8):946–949.

Crump KS [1984]. A new method for determining allowable daily intakes. Fund Appl Toxicol 4:854–871.

Crump KS, Howe R [1985]. A review of methods for calculating statistical confidence limits in low dose extrapolation. In: Clayson DB, Krewski D, Munro I, eds. Toxicological risk assessment. Vol. I. Boca Raton, FL: CRC Press, Inc.

Crystal RG, Gadek JE, Ferrans VJ, Fulmer JD, Line BR, Hunninghake GW [1981]. Interstitial lung disease: current concepts of pathogenesis, staging, and therapy. Am J Med 70:542–568.

Cullen RT, Jones AD, Miller BG, Tran CL, Davis JMG, Donaldson K, Wilson M, Stone V, Morgan A [2002]. Toxicity of volcanic ash from Montserrat. Edinburgh, UK: Institute of Occupational Medicine. IOM Research Report TM/02/01.

Dankovic D, Kuempel E, Wheeler M [2007]. An approach to risk assessment for TiO_2. Inhal Toxicol 19(Supp 1):205–212.

DFG [2000]. Deutsche Forschungsgemeinschaft. List of MAK and BAT values 2000. Weinheim, Germany: Wiley-VCH. Report No. 36, p. 102.

DFG [2008]. Deutsche Forschungsgemeinschaft. List of MAK and BAT values 2008. Weinheim, Germany: Wiley-VCH. Report No. 44, pp. 122, 137–143, II.

Dick CAJ, Brown DM, Donaldson K, Stone V [2003]. The role of free radicals in the toxic and inflammatory effects of four different ultrafine particle types. Inhal Toxicol 15(1):39–52.

DOI [2005]. Mineral commodity summaries 2005. Washington, DC: U.S. Department of the Interior, U.S. Geological Survey.

DOI [2008]. Titanium and titanium dioxide. In: Mineral commodity summaries 2008, p. 180–181. Washington, DC: U.S. Department of the Interior, U.S. Geological Survey [http://minerals.usgs.gov/minerals/pubs/commodity/titanium/mcs-2008-timet.pdf].

Donaldson K, Beswick PH, Gilmour PS [1996]. Free radical activity associated with the surface of particles: a unifying factor in determining biological activity? Toxicol Lett 88:293–298.

Donaldson K, Brown GM, Brown DM, Robertson MD, Slight J, Cowie H, Jones AD, Bolton AE, Davis JMG [1990]. Contrasting bronchoalveolar leukocyte responses in rats inhaling coal mine dust, quartz, or titanium dioxide: effects of coal rank, airborne mass concentration, and cessation of exposure. Environ Res 52:62–76.

Donaldson K, Stone V [2003]. Current hypotheses on the mechanisms of toxicity of ultrafine particles. Ann Ist Super Sanità 39(3):405–410.

Douglas AN, Robertson A, Chapman JS, Ruckley VA [1986]. Dust exposure, dust recovered from the lung, and associated pathology in a group of British coalminers. Br J Ind Med 43:795–801.

Driscoll KE [1996]. Role of inflammation in the development of rat lung tumors in response to chronic particle exposure. In: Mauderly JL, McCunney RJ, eds. Particle overload in the rat lung and lung cancer, implications for human risk assessment. Proceedings of the Massachusetts Institute of Technology Conference. Washington, DC: Taylor and Francis, pp.139–153.

Driscoll KE [2000]. TNFα and MIP-2: role in particle-induced inflammation and regulation by oxidative stress. Toxicol Letters 112–113:177–184.

Driscoll KE [2002]. E-mail message on October 29, 2002, from Kevin E. Driscoll, Proctor and Gamble Company, Cincinnati, Ohio, to Eileen Kuempel, Education and Information Division, National Institute for Occupational Safety and Health, Public Health Service, U.S. Department of Health and Human Services.

Driscoll KE, Carter JM, Howard BW, Hassenbein DG, Pepelko W, Baggs RB, Oberdörster G [1996]. Pulmonary inflammatory, chemokine, and mutagenic responses in rats after subchronic inhalation of carbon black. Toxicol Appl Pharmacol 136:372–380.

Driscoll KE, Deyo LC, Carter JM, Howard BW, Hassenbein DG, Bertram TA [1997]. Effects of particle exposure and particle-elicited inflammatory cells on mutation in rat alveolar epithelial cells. Carcinogenesis 18(2):423–430.

Driscoll KE, Lindenschmidt RC, Maurer JK, Higgins JM, Ridder G [1990]. Pulmonary response to silica or titanium dioxide: inflammatory cells, alveolar macrophage-derived cytokines, and histopathology. Am J Respir Cell Mol Biol 2(4):381–390.

Driscoll KE, Lindenschmidt RC, Maurer JK, Perkins L, Perkins M, Higgins J [1991]. Pulmonary response to inhaled silica or titanium dioxide. Toxicol Appl Pharmacol 111:201–210.

Duffin R, Tran L, Brown D, Stone V, Donaldson K [2007]. Proinflammogenic effects of low-toxicity and metal nanoparticles in vivo and in vitro: highlighting the role of particle surface area and surface reactivity. Inhal Toxicol 19(10):849–856.

Dunford R, Salinaro A, Cai L, Serpone N, Horikoshi S, Hidaka H, Knowland J [1997]. Chemical oxidation and DNA damage catalysed by inorganic sunscreen ingredients. FEBS Lett 418(1–2):87–90.

DuPont [2007]. Nanomaterial risk assessment worksheet DuPont™ light stabilizer [http://www.edf.org/documents/6913_TiO$_2$_worksheet.pdf].

Efron B, Tibshirani RJ [1998]. An introduction to the bootstrap. New York: Chapman & Hall, International Thomson Publishing.

Egerton TA [1997]. Titanium compounds (inorganic). In: Kroschwitz JI, Howe-Grant, eds. Kirk-Othmer encyclopedia of chemical technology. 4th ed. Vol. 24. New York: John Wiley & Sons, pp. 235–250.

Elder A, Gelein R, Finkelstein JN, Driscoll KE, Harkema J, Oberdörster G [2005]. Effects of subchronically inhaled carbon black in three species. I. Retention kinetics, lung inflammation, and histopathology. Toxicol Sci 88(2):614–629.

Elo R, Määttä K, Uksila E, Arstila AU [1972]. Pulmonary deposits of titanium dioxide in man. Arch Path 94:417–424.

EPA [1994]. Methods for derivation of inhalation reference concentrations and application of inhalation dosimetry. Washington, DC: U.S. Environmental Protection Agency, Office of Research and Development, EPA/600/8-90/066F [http://nepis.epa.gov/EPA/html/Pubs/pubtitleORD.htm].

EPA [2003]. Benchmark dose software, Version 1.3.2. Washington, DC: U.S. Environmental Protection Agency, National Center for Environmental Assessment.

EPA [2007]. Benchmark dose software, Version 1.4.1b. Washington, DC: U.S. Environmental Protection Agency, National Center for Environmental Assessment.

Everitt JI, Mangum JB, Bermudez E, Wong BA, Asgharian B, Reverdy EE, Hext PM, Warheit DB [2000]. Comparison of selected pulmonary responses of rats, mice and Syrian golden hamsters to inhaled pigmentary titanium dioxide. Inhal Toxicol. 12(Suppl 3):275–282.

Everitt JI, Preston RJ [1999]. Carcinogenicity and genotoxicity of inhaled substances. Chapter 10. In: Gardner DE, Crapo JD, McClellan RO, eds. Toxicology of the Lung. 3rd ed. Philadelphia, PA: Taylor & Francis, pp. 269–288.

Fayerweather WE, Karns ME, Gilby PG, Chen JL [1992]. Epidemiologic study of lung cancer mortality in workers exposed to titanium tetrachloride. J Occup Med 34(2):164–169.

Ferin J, Oberdörster G, Penney DP [1992]. Pulmonary retention of ultrafine and fine particles in rats. Am J Respir Cell Mol Biol 6:535–542.

Freedman AP, Robinson SE [1988]. Noninvasive magnetopneumographic studies of lung dust retention and clearance in coal miners. In: Frantz RL, Ramani RV, eds. Respirable dust in the mineral industries: health effects, characterization, and control. University Park, PA: The Pennsylvania State University, pp. 181–186.

Fryzek JP [2004]. E-mails during 2004, from Jon Fryzek to Chris Sofge, Education and Information Division, National Institute for Occupational Safety and Health, Centers for Disease Control and Prevention, Public Health Service, U.S. Department of Health and Human Services.

Fryzek JP, Chadda B, Marano D, White K, Schweitzer S, McLaughlin JK, Blot WJ [2003]. A cohort mortality study among titanium dioxide manufacturing workers in the United States. J Occup Environ Med 45:400–409.

Fryzek JP, Cohen S, Chadda B, Marano D, White K, McLaughlin JK, Blot WJ [2004a]. Errata: RE Fryzek et al., August 2004. J Occup Environ Med 46(11):1189.

Fryzek JP, Cohen S, Chadda B, Marano D, White K, McLaughlin JK, Blot WJ [2004b]. Titanium dioxide and lung cancer. [letter to the editor]. J Occup Environ Med 46(8):760.

Garabrant DH, Fine LJ, Oliver C, Bernstein L, Peters JM [1987]. Abnormalities of pulmonary function and pleural disease among titanium metal production workers. Scand J Work Environ Health 13:47–51.

Gaylor D, Ryan L, Krewski D, Zhu Y [1998]. Procedures for calculating benchmark doses for health risk assessment. Regul Toxicol Pharm 28:150–164.

Geiser M, Casaulta M, Kupferschmid B, Schulz H, Semmler-Behnke M, Kreyling W [2008]. The role of macrophages in the clearance of inhaled ultrafine titanium dioxide particles. Am J Respir Cell Mol Biol 38:371–376.

Geiser M, Rothen-Rutishauser B, Kapp N, Schurch S, Kreyling W, Schulz H, Semmler M, Im Hof V, Heyder J, Gehr P [2005]. Ultrafine particles cross cellular membranes by nonphagocytic mechanisms in lungs and in cultured cells. Environ Health Perspect 113(11):1555–1560.

Goodman GB, Kaplan PD, Stachura I, Castranova V, Pailes WH, Lapp NL [1992]. Acute silicosis responding to corticosteroid therapy. Chest 101:366–370.

Grassian VH, O'Shaughnessy PT, Adamcakova-Dodd A, Pettibone JM, Thorne PS [2007]. Inhalation exposure study of titanium dioxide nanoparticles with a primary particle size of 2 to 5 nm. Environ Health Perspec 115(3):397–402.

Green FHY, Vallyathan V, Hahn FF [2007]. Comparative pathology of environmental lung disease: an overview. Toxicologic Pathology 35:136–147.

Greim H, Ziegler-Skylakakis K [2007]. Risk assessment for biopersistent granular particles. Inhal Toxicol 19(Suppl 1):199–204.

Güngör N, Godschalk RWL, Pachen DM, Van Schooten FR, Knaapen AM [2007]. Activated neutrophils inhibit nucleotide excision repair in human pulmonary epithelial cells: role of myeloperoxidase. FASEB J 21:2359–2367 [http://www.fasebj.org/cgi/content/abstract/21/10/2359].

Gurr J-R, Wang ASS, Chen C-H, Jan K-Y [2005]. Ultrafine titanium dioxide particles in the absence of photoactivation can induce oxidative damage to human bronchial epithelial cells. Toxicology 213:66–73.

Haslam PL, Demwar A, Butchers P, Primett ZS, Newman-Taylor A, Turner-Warwick M [1987]. Mast cells, athpical lymphocytes, and neutrophils in bronchoalveolar lavage in extrinsic allergic alveolitis. Am Rev Respir Dis 135:35–47.

Heinrich U [1996]. Comparative response to long-term particle exposure among rats, mice, and hamsters. In: Mauderly JL, McCunney RJ, eds. Particle overload in the rat lung and lung cancer, implications for human risk assessment. Proceedings of the Massachusetts Institute of Technology Conference. Washington, DC: Taylor and Francis, pp. 51–71.

Heinrich U, Fuhst R, Rittinghausen S, Creutzenberg O, Bellmann B, Koch W, Levsen K [1995]. Chronic inhalation exposure of Wistar rats and two different strains of mice to diesel-engine exhaust, carbon black, and titanium dioxide. Inhal Toxicol 7(4):533–556.

Henderson RF, Driscoll KE, Harkema JR, Lindenschmidt RC, Chang I-Y, Maples KR, Barr EB [1995]. A comparison of the inflammatory response of the lung to inhaled versus instilled particles in F344 rats. Fund Appl Toxicol 24:183–197.

Hext PM, Tomenson JA, Thompson P [2005]. Titanium dioxide: inhalation toxicology and epidemiology. Ann Occup Hyg 49(6):461–472.

Hinds WC [1999]. Aerosol technology: properties, behavior, and measurement of airborne particles. 2nd ed. New York: John Wiley & Sons.

Höhr D, Steinfartz Y, Schins RPF, Knaapen AM, Martra G, Fubini B, Borm PJA [2002]. The surface area rather than the surface coating determines the acute inflammatory response after instillation of fine and ultrafine TiO$_2$ in the rat. Int J Hyg Environ Health 205:239–244.

Hseih TH, Yu CP [1998]. Two-phase pulmonary clearance of insoluble particles in mammalian species. Inhal Toxicol 10(2):121–130.

Huang S-H, Hubbs AF, Stanley CF, Vallyathan, V, Schnabel PC, Rojanasakul Y, Ma JKH, Banks DE, Weissman DN [2001]. Immunoglobulin responses to experimental silicosis. Toxicol Sci 59:108–117.

Hubbard AK, Timblin CR, Shukla A, Rincón M, Mossman BT [2002]. Activation of NF-6B-dependent gene expression by silica in lungs of luciferase reporter mice. Am J Physiol Lung Cell Mol Physiol 282:L968–L975.

Hudson DJ [1966]. Fitting segmented curves whose join points have to be estimated. J Amer Statistic Assoc 61(316):1097–1129.

Hughes DA, Haslam PL [1990]. Effect of smoking on the lipid composition of lung lining fluid and relationship between immunostimulatory lipids, inflammatory cells and foamy macrophages in extrinsic allergic alveolitis. Eur Respir J 3:1128–1139.

Humble S, Tucker JA, Boudreaux C, King JAC, Snell K [2003]. Titanium particles identified by energy-dispersive X-ray microanalysis within the lungs of a painter at autopsy. Ultrastruct Pathol 27:127–129.

IARC [1989]. IARC monographs on the evaluation of carcinogenic risks to humans: some organic solvents, resin monomers and related compounds, pigments and occupational exposures in paint manufacture and painting. Vol. 47. Lyon, France: World Health Organization, International Agency for Research on Cancer.

IARC [2010]. IARC monographs on the evaluation of carcinogenic risks to humans: carbon black, titanium dioxide, and talc. Vol. 93. Lyon, France: World Health Organization, International Agency for Research on Cancer. [http://monographs.iarc.fr/ENG/Monographs/vol93/index.php].

ICRP [1994]. Human respiratory tract model for radiological protection. In: Smith H, ed. Annals of the ICRP. Tarrytown, New York: International Commission on Radiological Protection, ICRP Publication No. 66.

ILSI (International Life Sciences Institute) [2000]. The relevance of the rat lung response to particle overload for human risk assessment: a workshop consensus report. Inhal Toxicol 12:1–17.

ISO [1995]. Air quality—particle size fraction definitions for health-related sampling. Geneva, Switzerland: International Organization for Standardization, ISO Report No. ISO 7708.

Janssen YM, Marsh JP, Driscoll KE, Borm PJ, Oberdörster G, Mossman BT [1994]. Increased expression of manganese-containing superoxide dismutase in rat lungs after inhalation of inflammatory and fibrogenic minerals. Free Radic Biol Med 16(3):315–322.

Jiang J, Oberdörster G, Elder A, Gelein R, Mercer P, Biswas P [2008]. Dose nanoparticle activity depend upon size and crystal phase? Nanotox 2(1):33–42.

Kakinoki K, Yamane K, Teraoka R, Otsuka M, Matsuda Y [2004]. Effect of relative humidity on the photocatalytic activity of titanium dioxide and photostability of famotidine. J Pharm Sci 93(3):582–589.

Kang JL, Moon C, Lee HS, Lee HW, Park EM, Kim HS, Castranova V [2008]. Comparison of the biological activity between ultrafine and fine titanium dioxide particles in RAW 264.7 cells associated with oxidative stress. J Toxicol Environ Health A 71(8):478–485.

Katabami M, Dosaka-Akita H, Honma K, Saitoh Y, Kimura K, Uchida Y, Mikami H, Ohsaki Y, Kawakami Y, Kikuchi K [2000]. Pneumoconiosis-related lung cancers: preferential occurrence from diffuse interstitial fibrosis-type pneumoconiosis. Am J Respir Crit Care Med 162:295–300.

Kawahara T, Ozawa T, Iwasaki M, Tada H, Ito S [2003]. Photocatalytic activity of rutile-anatase coupled TiO_2 particles prepared by a dissolution-reprecipitation method. J Colloid Interface Sci 267(2):377–381.

Keller CA, Frost A, Cagle PT, Abraham JL [1995]. Pulmonary alveolar proteinosis in a painter with elevated pulmonary concentrations of titanium. Chest 108:277–280.

Knaapen AM, Albrecht C, Becker A, Höhr D, Winzer A, Haenen GR, Borm PJA, Shins RPF [2002]. DNA damage in lung epithelial cells isolated from rats exposed to quartz: role of surface reactivity and neutrophilic inflammation. Carcinogenesis 23(7):1111–1120.

Knaapen AM, Borm PJA, Albrecht C, Shins RPF [2004]. Inhaled particles and lung cancer. Part A: Mechanisms. Int J Cancer 109:799–809.

Kuempel ED, Smith RJ, Dankovic DA, Bailer AJ, Stayner LT [2002]. Concordance of rat and human based risk estimates for particle-related lung cancer. Ann Occup Hyg 46(Suppl 1):62–66.

Kuempel ED, O'Flaherty EJ, Stayner LT, Smith RJ, Green FHY, Vallyathan V [2001a]. A biomathematical model of particle clearance and retention in the lungs of coal miners. Part I. Model development. Reg Toxicol Pharmacol 34:69–87.

Kuempel ED, Tran CL, Smith RJ, Bailer AJ [2001b]. A biomathematical model of particle clearance and retention in the lungs of coal miners. Part II. Evaluation of variability and uncertainty. Reg Toxicol Pharmacol 34:88–101.

Kuempel ED, Tran CL, Bailer AJ, Porter DW, Hubbs AF, Castranova V [2001c]. Biological and statistical approaches to predicting human lung cancer risk from silica. J Environ Pathol Toxicol Oncol 20(Suppl 1):15–32.

Kuschner M [1995]. The relevance of rodent tumors in assessing carcinogenicity in human beings. Regul Toxicol Pharmacol 21(2):250–251.

Lapp NL, Castranova V [1993]. How silicosis and coal workers' pneumoconiosis develop: a cellular assessment. State Art Rev Occup Med 8(1):35–56.

Lauweryns JM, Baert JH [1977]. Alveolar clearance and the role of the pulmonary lymphatics. Am Rev Respir Dis 115:625–683.

Le Bouffant L [1971]. Influence de la nature des poussieres et de la charge pulmonaire sur l'epuration. In: Walton WH, ed. Proceedings of inhaled particles, III, vol. 1. Oxford, UK: Pergamon Press, pp. 227–237.

Lee KP, Henry NW III, Trochimowicz HJ, Reinhardt CF [1986a]. Pulmonary response to impaired lung clearance in rats following excessive TiO_2 dust deposition. Environ Res 41:144–167.

Lee KP, Kelly DP, Schneider PW, Trochimowicz HJ [1986b]. Inhalation toxicity study on rats exposed to titanium tetrachloride atmospheric hydrolysis products for two years. Toxicol Appl Pharmacol 83:30–45.

Lee KP, Trochimowicz HJ, Reinhardt CF [1985]. Pulmonary response of rats exposed to titanium dioxide (TiO$_2$) by inhalation for two years. Toxicol Appl Pharmacol 79:179–192.

Lee KW, Liu BYH [1981]. Experimental study of aerosol filtration by fibrous filters. Aerosol Sci Technol 1(1):35–46.

Lee KW, Liu BYH [1982]. Theoretical study of aerosol filtration by fibrous filters. Aerosol Sci Technol 1(2):147–161.

Lehnert BE [1993]. Defense mechanisms against inhaled particles and associated particle-cell interactions. In: Guthrie and Mossman, eds. Health effects of mineral dusts. Vol. 28. Washington, DC: Mineralogical Society of America, pp. 427–469.

Levy LS [1996]. Differences between rodents and humans in lung tumor response—lessons from recent studies with carbon black. In: Mauderly JL, McCunney RJ, eds. Particle overload in the rat lung and lung cancer, implications for human risk assessment. Proceedings of the Massachusetts Institute of Technology Conference. Washington, DC: Taylor and Francis, pp. 125–138.

Lewis RJ Sr. [1993]. Hawley's condensed chemical dictionary. 12th ed. New York:Van Nostrand Reinhold Company, p. 1153.

Lewis TR, Green FHY, Moorman WJ, Burg JR, Lynch DW [1989]. A chronic inhalation toxicity study of diesel engine emissions and coal dust, alone and combined. J Amer Coll Toxicol 8(2):345–375.

Lewis TR, Morrow PE, McClellan RO, Raabe OG, Kennedy GL, Schwetz BA, Goehl TJ, Roycroft JH, Chhabra RS [1989]. Establishing aerosol exposure concentrations for inhalation toxicity studies. Toxicol Appl Pharmacol 99:377–383.

Litovitz T [2004]. E-mail message on June 23, 2004, from T Litovitz to AF Hubbs, Health Effects Laboratory Division, National Institute for Occupational Safety and Health, Centers for Disease Control and Prevention, Public Health Service, U.S. Department of Health and Human Services.

Litovitz TL, Klein-Schwartz W, Rodgers GC, Cobaugh DJ, Youniss J, Omslaer JC, May ME, Woolf AD, Benson BE [2002]. 2001 Annual report of the American Association of Poison Control Centers toxic exposure surveillance system. Am J Emerg Med 20(5):391–452.

Long TC, Tajuba J, Sama P, Saleh N, Swartz C, Parker J, Hester S, Lowry GV, Veronesi B [2007]. Nanosize titanium dioxide stimulates reactive oxygen species in brain microglia and damages neurons in vitro. Environ Health Perspect 115(11):1631–1637.

Lu P-J, Ho I-C, Lee T-C [1998]. Induction of sister chromatid exchanges and micronuclei by titanium dioxide in Chinese hamster ovary-K1 cells. Mutat Res 414(1–3):15–20.

Määttä K, Arstila AU [1975]. Pulmonary deposits of titanium dioxide in cytologic and lung biopsy specimens: light and electron microscopic X-ray analysis. Lab Invest 33(3):342–346.

Maier M, Hannebauer B, Holldorff H, Albers P [2006]. Does lung surfactant promote disaggregation of nanostructured titanium dioxide? J Occup Environ Med 48(12):1314–1320.

Malvern Instruments [2004]. Measurements of TiO$_2$ particle size distribution. Southborough, MA: Malvern Instruments. Contract for American Chemistry Council.

Maronpot RR, Flake G, Huff J [2004]. Relevance of animal carcinogenesis findings to human cancer predictions and prevention. Toxicol Pathol 32(Suppl 1):40–48.

Martin JC, Dániel H, LeBouffant L [1977]. Short- and long-term experimental study of the toxicity of coal mine dust and of some of its constituents. In: Walton WH, ed. Inhaled particles, Part l. 4th ed. Oxford, UK: Pergamon Press, 361–370.

Martin TR, Ganesh R, Maunder RJ, Springmeyer SC [1985]. The effects of chronic bronchitis and chronic air-flow obstruction on lung cell populations recovered by bronchoalveolar lavage. Am Rev Respir Dis 132:254–260.

Mauderly JL [1996]. Lung overload: the dilemma and opportunities for resolution. Introduction to lung overload. In: Mauderly JL, McCunney RJ, eds. Particle overload in the rat lung and lung cancer, implications for human risk assessment. Proceedings of the Massachusetts Institute of Technology Conference. Washington, DC: Taylor and Francis, pp. 1–28.

Mauderly JL [1997]. Relevance of particle-induced rat lung tumors for assessing lung carcinogenic hazard and human lung cancer risk. Environ Health Perspect 105(Suppl 5):1337–1346.

Mauderly JL, Jones RK, Griffith WC, Henderson RF, McClellan RO [1987]. Diesel exhaust is a pulmonary carcinogen in rats exposed chronically by inhalation. Fundam Appl Toxicol 9:208–221.

Mercer RR, Russell ML, Roggli VL, Crapo JD [1994]. Cell number and distribution in human and rat airways. Am J Respir Cell Mol Biol 10:613–624.

Miller FJ [1999]. Dosimetry of particles in laboratory animals and humans. Chapter 18. In: Gardner DE, Crapo JD, McClellan RO, eds. Toxicology of the lung, 3rd ed. Philadelphia, PA: Taylor & Francis, pp. 513–556.

Mohr U, Ernst H, Roller M, Pott F [2006]. Pulmonary tumor types induced in Wistar rats of the so-called "19-dust study." Exp Toxicol Pathol 58(1):13–20.

Möller W, Hofer T, Ziesenis A, Karg E, Heyder J [2002]. Ultrafine particles cause cytoskeletal dysfunctions in macrophages. Toxicol Appl Pharmacol 182(3):197–207.

Moran CA, Mullick FG, Ishak KG, Johnson FB, Hummer WB [1991]. Identification of titanium in human tissues: probable role in pathologic processes. Hum Pathol 22(5):450–454.

Morfeld P, Albrecht C, Drommer W, Borm PJA [2006]. Dose-response and threshold analysis of tumor prevalence after intratracheal instillation of six types of low- and high-surface-area particles in a chronic rat experiment. Inhal Toxicol 18(4):215–225.

Morrow PE [1988]. Possible mechanisms to explain dust overloading of the lungs. Fund Appl Toxicol 10:369–384.

Morrow PE, Muhle H, Mermelstein R [1991]. Chronic inhalation study findings as a basis for proposing a new occupational dust exposure limit. J Amer College Toxicol 10(2):279–290.

Muhle H, Bellmann B, Creutzenberg O [1994]. Toxicokinetics of solid particles in chronic rat studies using diesel soot, carbon black, toner, titanium dioxide, and quartz. In: Dungworth DL, Mauderly JL, Oberdörster G, eds. Toxic and carcinogenic effects of solid particles in the respiratory tract. ILSI Monographs. Washington, DC: International Life Sciences Institute/ILSI Press, pp. 29–41.

Muhle H, Bellmann B, Creutzenberg O, Dasenbrock C, Ernst H, Kilpper R, MacKenzie JC, Morrow P, Mohr U, Takenaka S, Mermelstein R [1991]. Pulmonary response to toner upon chronic inhalation exposure in rats. Fund Appl Toxicol *17*:280–299.

Muhle H, Takenaka S, Mohr U, Dasenbrock C, Mermelstein R [1989]. Lung tumor induction upon long-term, low-level inhalation of crystalline silica. Am J Ind Med *15*:343–346.

Myhr BC, Caspary WJ [1991]. Chemical mutagenesis at the thymidine kinase locus in L5178Y mouse lymphoma cells: results for 31 coded compounds in the National Toxicology Program. Environ Mol Mutagen *18*(1):51–83.

Nakagawa Y, Wakuri S, Sakamoto K, Tanaka N [1997]. The photogenotoxicity of titanium dioxide particles. Mutat Res *394*:125–132.

NCHS [1996]. Vital statistics of the United States, 1992. Vol. II. Mortality. Part A. Hyattsville, MD: U.S. Department of Health and Human Services, Public Health Service, Centers for Disease Control and Prevention, National Center for Health Statistics, Tables 1–27, 6–2, 7–2.

NCI [1979]. Bioassay of titanium dioxide for possible carcinogenicity. Washington, DC: U.S. Department of Health, Education, and Welfare, Public Health Service, National Institutes of Health, National Cancer Institute Technical Report Series No. 97.

Nemmar A, Melghit K, Ali BH [2008]. The acute proinflammatory and prothrombotic effects of pulmonary exposure to rutile TiO_2 nanorods in rats. Exp Biol Med *233*:610–619.

Nikula KJ, Avila KJ, Griffith WC, Mauderly JL [1997]. Lung tissue responses and sites of particle retention differ between rats and cynomolgus monkeys exposed chronically to diesel exhaust and coal dust. Fund Appl Toxicol *37*:37–53.

Nikula KJ, Snipes MB, Barr EB, Griffith WC, Henderson RF, Mauderly JL [1995]. Comparative pulmonary toxicities and carcinogenicities of chronically inhaled diesel exhaust and carbon black in F344 rats. Fundam Appl Toxicol *25*:80–94.

Nikula KJ, Vallyathan V, Green FHY, Hah FF [2001]. Influence of dose on the distribution of retained particulate material in rat and human lungs. In: Proceedings of Particulate Matter 2000, January 23–28, 2000. Charleston, SC: Air and Waste Management Association.

NIOSH [1995]. Criteria for a recommended standard: occupational exposure to respirable coal mine dust. Cincinnati, OH: U.S. Department of Health and Human Services, Public Health Service Centers for Disease Control and Prevention, National Institute for Occupational Safety and Health, DHHS (NIOSH) Publication No. 95–106.

NIOSH [1998]. Particulates not otherwise regulated, respirable. Method 0600 (supplement issued January 15, 1998). In: NIOSH manual of analytical methods. Cincinnati, OH: U.S. Department of Health and Human Services, Public Health Service Centers for Disease Control and Prevention, National Institute for Occupational Safety and Health, DHHS (NIOSH) Publication No. 94–113 [http://www.cdc.gov/niosh/nmam/pdfs/0600.pdf].

NIOSH [2002]. NIOSH pocket guide to chemical hazards and other databases. CD-ROM. Cincinnati, OH: U.S. Department of Health and Human Services, Public Health Service, Centers for Disease Control, National Institute for Occupational Safety and Health, DHHS (NIOSH) Publication No. 2002–140.

NIOSH [2003]. Elements by ICP (Nitric/Perchloric acid ashing). Method 7300 (supplement issued Mary 15, 2003). In: NIOSH manual of analytical methods, 4th ed. Cincinnati, OH: U.S. Department of Health and Human Services, Public Health Service, Centers for Disease Control and Prevention, National Institute for Occupational Safety and Health, DHHS (NIOSH) Publication 94-113 [http://www.cdc.gov/niosh/nmam/pdfs/7300.pdf].

NIOSH [2004]. NIOSH respirator selection logic. Cincinnati, OH: U.S. Department of Health and Human Services, Public Health Service, Centers for Disease Control and Prevention, National Institute for Occupational Safety and Health, DHHS (NIOSH) Publication No. 2005-100 [www.cdc.gov/niosh/docs/2005-100/].

NIOSH [2006]. NIOSH criteria for a recommended standard: occupational exposure to refractory ceramic fibers. Cincinnati, OH: U.S. Department of Health and Human Services, Centers for Disease Control and Prevention, National Institute for Occupational Safety and Health, DHHS (NIOSH) Publication No. 2006-123.

NIOSH [2009a]. Approaches to safe nanotechnology: managing the health and safety concerns associated with engineered nanomaterials. Cincinnati, OH: U.S. Department of Health and Human Services, Public Health Service, Centers for Disease Control and Prevention, National Institute for Occupational Safety and Health, DHHS (NIOSH) Publication No. 2009-125 [http://www.cdc.gov/niosh/topics/nanotech/safenano/].

NIOSH [2009b]. Health Hazard Evaluation Report: evaluation of employees' exposures to welding fumes and powder paint dust during metal furniture manufacturing. By Rodriguez M, Adebayo A, Brueck S, Ramsey, J. Cincinnati, OH: U.S. Department of Health and Human Services, Centers for Disease Control and Prevention, National Institute for Occupational Safety and Health, NIOSH HETA No. 2007-0199-3075.

NIOSH [2009c]. Current intelligence bulletin 60: interim guidance for medical screening and hazard surveillance for workers potentially exposed to engineered nanoparticles. By Schulte PS, Trout D, Zumwalde RD. Cincinnati, OH: U.S. Department of Health and Human Services, Centers for Disease Control and Prevention, National Institute for Occupational Safety and Health, DHHS (NIOSH) Publication 2009-116.

NTP [1993]. Toxicology and carcinogenesis studies of talc (CAS N0. 14807-96-6) in F344/N Rats and B6C3F$_1$ mice (inhalation studies). Research Triangle Park, NC: U.S. Department of Health and Human Services, Public Health Service, National Institutes of Health, National Toxicology Program, NTP Technical Report 421 [http://ntp.niehs.nih.gov/ntp/htdocs/lt_rpts/tr421.pdf].

Nurkiewicz TR, Porter DW, Hubbs AF, Cumpston JL, Chen BT, Frazer DG, Castranova V [2008]. Nanoparticle inhalation augments particle-dependent systemic microvascular dysfunction. Part Fibre Toxicol 5:1.

Oberdörster G [1996]. Significance of particle parameters in the evaluation of exposure-dose-response relationships of inhaled particles. In: Mauderly JL, McCunney RJ, eds. Particle overload in the rat lung and lung cancer, implications for human risk assessment. Proceedings of the Massachusetts Institute of Technology Conference. Washington, DC: Taylor and Francis, pp.139-153.

Oberdörster G, Ferin J, Gelein F, Soderholm SC, Finkelstein J [1992]. Role of the alveolar macrophage in lung injury: studies with ultrafine particles. Environ Health Perspect 97:193-199.

Oberdörster G, Ferin J, Lehnert BE [1994a]. Correlation between particle size, *in vivo* particle persistence, and lung injury. Environ Health Perspect *102*(Suppl 5):173–179.

Oberdörster G, Ferin J, Soderholm S, Gelein R, Cox C, Baggs R, Morrow PE [1994b]. Increased pulmonary toxicity of inhaled ultrafine particles: due to lung overload alone? Ann Occup Hyg *38*:295–302.

Oberdörster G, Yu CP [1990]. The carcinogenic potential of inhaled diesel exhaust: a particle effect? J Aerosol Sci *21*(Suppl 1):S397–S401.

Ophus EM, Rode L, Gylseth B, Nicholson DG, Saeed K [1979]. Analysis of titanium pigments in human lung tissue. Scand J Work Environ Health *5*:290–296.

OSHA [2002]. Metal and metalloid particulates in workplace atmospheres (atomic absorption). Washington, DC: U.S. Department of Labor, Occupational Safety and Health Administration [http://www.osha.gov/dts/sltc/methods/inorganic/id121/id121.html].

Park R, Rice F, Stayner L, Smith R, Gilbert S, Checkoway H [2002]. Exposure to crystalline silica, silicosis, and lung disease other than cancer in diatomaceous earth industry workers: a quantitative risk assessment. Occup Environ Med *59*:36–43.

Porter DW, Ramsey DM, Hubbs AF, Battelli L, Ma JYC, Burger M, Landsittel D, Robinson VA, McLaurin JL, Khan A, Jones W, Teass A, Castranova V [2001]. Time course of pulmonary response of rats to inhalation of crystalline silica: histological results are biochemical indices of damage, lipidosis, and fibrosis. J Environ Pathol Toxicol Oncol *20*(Suppl 1):1–14.

Pott F, Roller M [2005]. Carcinogenicity study with nineteen granular dusts in rats. Eur J Oncol *10*(4):249–281.

Pott F, Ziem U, Reiffer F-J, Huth, F, Ernst H, Mohr U [1987]. Carcinogenicity studies on fibres, metal compounds and some other dusts in rats. Exp Pathol *32*:129–152.

Rahman Q, Lohani M, Dopp E, Pemsel H, Jonas L, Weiss DG, Schiffmann D [2002]. Evidence that ultrafine titanium dioxide induces micronuclei and apoptosis in Syrian hamster embryo fibroblasts. Environ Health Perspect *110*(8):797–800.

Ramanakumar AV, Parent ME, Latreille B, Siemiatycki J [2008]. Risk of lung cancer following exposure to carbon black, titanium dioxide and talc: results from two case-control studies in Montreal. Int J Cancer *122*(1):183–189.

Raterman SM [1996]. Methods of control. Chapter 18. In: Plog BA, ed. Fundamentals of industrial hygiene. Itasca, IL: National Safety Council.

Rehn B, Seiler F, Rehn S, Bruch J, Maier M [2003]. Investigations on the inflammatory and genotoxic lung effects of two types of TiO_2: untreated and surface treated. Toxicol Appl Pharmacol *189*(2):84–95.

Rengasamy S, King WP, Eimer B, Shaffer RE [2008]. Filtration performance of NIOSH-approved N95 and P100 filtering-facepiece respirators against 4–30 nanometer size nanoparticles. J Occup Environ Hyg *5*(9):556–564.

Rengasamy S, Verbofsky R, King WP, Shaffer RE [2007]. Nanoparticle penetration through NIOSH-approved N95 filtering-facepiece respirators. J Int Soc Respir Pro *24*:49–59.

Renwick LC, Brown D, Clouter A, Donaldson K [2004]. Increased inflammation and altered macrophage chemotactic responses caused by two ultrafine particles. Occup Environ Med *61*:442–447.

Renwick LC, Donaldson K, Clouter A [2001]. Impairment of alveolar macrophage phagocytosis by ultrafine particles. Toxicol Appl Pharmacol 172(2):119–127.

Rice FL, Park R, Stayner L, Smith R, Gilbert S, Checkoway H [2001]. Crystalline silica exposure and lung cancer mortality in diatomaceous earth industry workers: a quantitative risk assessment. Occup Environ Med 58(1):38–45.

Rittinghausen S, Mohr U, Dungworth DL [1997]. Pulmonary cystic keratinizing squamous cell lesions of rats after inhalation/instillation of different particles. Exp Toxicol Pathol 49:433–446.

Rode LE, Ophus EM, Glyseth B [1981]. Massive pulmonary deposition of rutile after titanium dioxide exposure. Acta Path Microbiol Scand, Sect A, 89:455–461.

Roller M [2007]. Differences between the data bases, statistical analyses, and interpretations of lung tumors of the 19-dust study: two controversial views. Exp Toxicol Pathol 58(6):393–405.

Roller M, Pott F [2006]. Lung tumor risk estimates from rat studies with not specifically toxic granular dusts. Ann NY Acad Sci 1076:266–280.

Rom WL [1991]. Relationship of inflammatory cell cytokines to disease severity in individuals with occupational inorganic dust exposure. Am J Indus Med 19:15–27.

Sanderson BJS, Wang JJ, Wang H [2007]. Reply to Letter to the Editor. Mutat Res 634:243–244.

Sayes CM, Wahi R, Kurian PA, Liu Y, West JL, Ausman KD, Warheit DB, Colvin VL [2006]. Correlating nanoscale titania structure with toxicity: a cytotoxicity and inflammatory response study with human dermal fibroblasts and human lung epithelial cells. Toxicol Sci 92(1):174–185.

Schins RPF, Knaapen AM [2007]. Genotoxicity of poorly soluble particles. Inhal Toxicol 19(Suppl. 1):189–198.

Seinfeld JA, Pandis SN [1998]. Atmospheric chemistry and physics. New York: John Wiley & Sons, Inc.

Shi X, Castranova V, Halliwell B, Vallyathan V [1998]. Reactive oxygen species and silica-induced carcinogenesis. J Toxicol Environ Health, Part B, 1:181–197.

Siemiatycki J, Bégin D, Dewar R, Gérin M, Lakhani R, Nadon L, Richardson L [1991]. Risk factors for cancer in the workplace. Boca Raton, FL: CRC Press, pp. 63, 153, 185, 272, 280.

Slade R, Crissman K, Norwood J, Hatch G [1993]. Comparison of antioxidant substances in bronchoalveolar lavage cells and fluid from humans, guinea pigs, and rats. Exp Lung Res 19(4):469–484.

Slade R, Stead AG, Graham JA, Hatch GE [1985]. Comparison of lung antioxidant levels in humans and laboratory animals. Am Rev Respir Dis 131(5):742–746.

Sleeuwenhoek A [2005]. Summary of occupational exposure measurements associated with production of titanium dioxide. Report No. 899-00055. Edinburgh, UK: Institute of Occupational Medicine.

Snipes MB [1996]. Current information on lung overload in nonrodent mammals: contrast with rats. In: Mauderly JL, McCunney RJ, eds. Particle overload in the rat lung and lung cancer, implications for human risk assessment. Proceedings of the Massachusetts Institute of Technology Conference. Washington, DC: Taylor and Francis, pp. 91–109.

Stayner L, Dankovic D, Smith R, Steenland K [1998]. Predicted lung cancer risk among miners exposed to diesel exhaust particles. Am J Ind Med 34(3):207–219.

Stayner LT, Dankovic DA, Smith RJ, Gilbert SJ, Bailer AJ [2000]. Human cancer risk and exposure to 1,3-butadiene—a tale of mice and men. Scand J Work Environ Health 26(4):322–330.

Stenbäck F, Rowland J, Sellakumar A [1976]. Carcinogenicity of benzo(a)pyrene and dusts in hamster lung (instilled intratracheally with titanium oxide, aluminum oxide, carbon, and ferric oxide). Oncology 33:29–34.

Stöber W, Einbrodt HJ, Klosterkötter W [1965]. Quantitative studies of dust retention animal and human lungs after chronic inhalation. In: Davis CN, ed. Proceedings of an International Symposium, Cambridge. New York: Pergamon Press, pp. 409–418.

Tran CL, Buchanan D [2000]. Development of a biomathematical lung model to describe the exposure-dose relationship for inhaled dust among U.K. coal miners. Edinburgh, UK: Institute of Occupational Medicine, IOM Research Report TM/00/02.

Tran CL, Buchanan D, Cullen RT, Searl A, Jones AD, Donaldson K [2000]. Inhalation of poorly soluble particles. II. Influence of particle surface area on inflammation and clearance. Inhal Toxicol 12(12):1113–1126.

Tran CL, Cullen RT, Buchanan D, Jones AD, Miller BG, Searl A, Davis JMG, Donaldson K [1999]. Investigation and prediction of pulmonary responses to dust. Part II. In: Investigations into the pulmonary effects of low toxicity dusts. Parts I and II. Suffolk, UK: Health and Safety Executive, Contract Research Report 216/1999.

Tripathy NK, Würgler FE, Frei H [1990]. Genetic toxicity of six carcinogens and six non-carcinogens in the Drosophila wing spot test. Mutat Res 242(3):169–180.

Türkez H, Geyikoğlu F [2007]. An in vitro blood culture for evaluating the genotoxicity of titanium dioxide: the responses of antioxidant enzymes. Toxicol Ind Health 23:19–23.

U.S. Supreme Court [1980]. Industrial Union Department, AFL-CIO v. American Petroleum Institute et al., Case Nos. 78-911, 78-1036. Supreme Court Reporter 100:2844–2905.

Vallyathan V, Goins M, Lapp LN, Pack D, Leonard S, Shi X, Castranova V [2000]. Changes in bronchoalveolar lavage indices associated with radiographic classification in coal miners. Am J Respir Crit Care Med 162:958–965.

Vallyathan V, Shi X, Castranova V [1998]. Reactive oxygen species: their relation to pneumoconiosis and carcinogenesis. Environ Health Perspect 106 (Suppl 5):1151–1155.

Vu V, Barrett JC, Roycroft J, Schuman L, Dankovic D, Bbaro P, Martonen T, Pepelko W, Lai D [1996]. Chronic inhalation toxicity and carcinogenicity testing of respirable fibrous particles. Workshop report. Regul Toxicol Pharmacol 24(3):202–212.

Wang JJ, Sanderson BJS, Wang H [2007a]. Cyto- and genotoxicity of ultrafine TiO_2 particles in cultured human lymphoblastoid cells. Mutat Res 628:99–106.

Wang J, Zhou G, Chen C, Yu H, Wang T, Ma Y, Jia G, Gao Y, Li B, Sun J, Li Y, Jiao F, Zhao Y, Chai Z [2007b]. Acute toxicity and biodistribution of different sized titanium dioxide particles in mice after oral administration. Toxicol Lett 168:176–185.

Warheit DB, Brock WJ, Lee KP, Webb TR, Reed KL [2005]. Comparative pulmonary toxicity inhalation and instillation studies with different TiO_2 particle formulations: impact of surface treatments on particle toxicity. Toxicol Sci *88*(2):514–524.

Warheit DB, Frame SR [2006]. Characterization and reclassification of titanium dioxide-related pulmonary lesions. J Occup Environ Med *48*:1308–1313.

Warheit DB, Hansen JF, Yuen IS, Kelly DP, Snajdr SI, Hartsky MA [1997]. Inhalation of high concentrations of low toxicity dusts in rats results in impaired pulmonary clearance mechanisms and persistent inflammation. Toxicol Appl Pharmacol *145*(1):10–22.

Warheit DB, Webb TR, Reed KL [2006a]. Pulmonary toxicity screening studies in male rats with TiO_2 particulates substantially encapsulated with pyrogenically deposited, amorphous silica. Particle Fibre Toxicol *3*:3 [published January 26, 2006].

Warheit DB, Webb TR, Sayes CM, Colvin VL, Reed KL [2006b]. Pulmonary instillation studies with nanoscale TiO_2 rods and dots in rats: toxicity is not dependent upon particle size and surface area. Toxicol Sci *91*(1):227–236.

Warheit DB, Webb TR, Reed KL, Frerichs S, Sayes CM [2007]. Pulmonary toxicity study in rats with three forms of ultrafine-TiO_2 particles: differential responses related to surface properties. Toxicology *230*:90–104.

Watson AY, Valberg PA [1996]. Particle-induced lung tumors in rats: evidence for species specificity in mechanisms. In: Mauderly JL, McCunney RK, eds. Particle overload in the rat lung and lung cancer. Implications for human risk assessment. Washington, DC: Taylor and Francis, pp. 227–257.

Wheeler MW, Bailer AJ [2007]. Properties of model-averaged BMDLs: a study of model averaging in dichotomous response risk estimation. Risk Anal *27*(3):659–670.

WHO [1994]. Environmental health criteria 170: assessing human health risks of chemicals: derivation of guidance values for health-based exposure limits. Geneva, Switzerland: World Health Organization, International Programme on Chemical Safety [http://www.inchem.org/documents/ehc/ehc/ehc170.htm].

Wicks ZW Jr. [1993]. Coatings. In: Kroschwitz JI, Howe-Grant, eds. Kirk-Othmer encyclopedia of chemical technology. 4th ed. Vol. 6. New York: John Wiley & Sons, pp. 669, 692–694, 746.

Xia T, Kovochich M, Brant J, Hotze M, Sempf J, Oberley T, Sioutas C, Yeh JI, Wiesner MR, Nel AE [2006]. Comparison of the abilities of ambient and manufactured nanoparticles to induce cellular toxicity according to an oxidative stress paradigm. Nano Lett *6*(8):1794–1807.

Yamadori I, Ohsumi S, Taguchi K [1986]. Titanium deposition and adenocarcinoma of the lung. Acta Pathol Jpn *36*(5):783–790.

APPENDIX A: Statistical Tests of the Rat Lung Tumor Models

A–1. Goodness of Fit of Models using Mass or Surface Area Dose Metric

As seen in Figures 3–4 and 3–5, particle surface area is a much better dose metric than particle mass for predicting the lung tumor response in rats after chronic exposure to fine and ultrafine TiO_2. The statistical fit of these models is shown in Table A–1, using either mass or particle surface area dose. These goodness-of-fit tests show that particle surface area dose provides an adequate fit to models using either the all tumor response or tumors excluding squamous cell keratinizing cysts and that particle mass dose provides an inadequate fit to these data. The P-values are for statistical tests of the lack of fit; thus, $P < 0.05$ indicates lack of fit.

A–2. Heterogeneity Tests (Rat Strain or Sex; Particle Size)

Because of the observed differences in tumor response in males and females, when squamous cell keratinizing cystic tumors were included in the analysis (Table 4–4), it was important to test for heterogeneity in response by rat sex. Since the data were from different studies and rat strains, these factors were also investigated for heterogeneity (the influence of study and strain could not be evaluated separately because a different strain was used in each study). Finally, the possibility of heterogeneity in response to fine and ultrafine TiO_2 after adjustment for particle surface area was investigated to determine whether other factors may be associated with particle size that influence lung tumor response and that may not have been accounted for by particle surface area dose. Table A–2 shows that there was statistically significant heterogeneity between male and female rats for the *all lung tumors* response but not for the tumors excluding squamous cell keratinizing cysts. No heterogeneity in tumor response was observed across study/strain or for fine versus ultrafine when dose was expressed as particle surface area per gram of lung. These analyses showed that all of the data from the different studies, rat strains, and both sexes [Lee et al. 1985; Muhle et al. 1991; Heinrich et al. 1995] could be pooled and the model fit was adequate when the dose metric used is particle surface area per gram of lung and the tumor response is neoplastic tumors (i.e., without squamous cell keratinizing cystic tumors).

In addition, a modified dose-response model was developed to examine the all-tumor response (by adjusting for rat sex and to include the averaged male/female lung tumor response data in the Muhle et al. [1991] study) (see Section A–3).

A–3. Modified Logistic Regression Model (All Tumors in Male and Female Rats)

A modified logistic regression model was constructed to use all tumor data (including squamous cell keratinizing cystic tumors) to account for heterogeneity in tumor response observed between male and female rats in the Lee et al. [1985] and Heinrich et al. [1995] studies. In addition, the Muhle et al. [1991] study reported tumor response for males and females combined. For these reasons, the standard models in the BMDS [EPA 2003] could not be used. The BMDS models do not allow for covariates (e.g., sex) or for alternative model structures to account for the combined data.

In the modified logistic regression model, the total tumor count was evaluated as the sum of tumors from two distinct binomial responses. This implies that the expected response can be modeled as

$$N_{obs} = n_m p_m + n_f p_f \quad \text{(equation 1)}$$

where $N = n_m + n_f$ and the set (p_m, p_f) are binomial probabilities of tumor response for males and females that are modeled using the same assumptions of logistic regression. For example, female rats would have the following response:

$$p_f = \frac{\exp(\alpha_f + \beta_f \cdot dose)}{1 + \exp(\alpha_f + \beta_f \cdot dose)} \quad \text{(equation 2)}$$

that is the same as a logistic model that investigates only female rats. Thus, to model responses across studies using male, female, and male/female combinations, equations (1) and (2) can be used when n_m and n_f are known. When they are not known (using results reported in Muhle et al. [1991]), these quantities are estimated to be $n/2$.

With p_m and p_f now estimable using all data, the benchmark dose (BMD) can be computed by methods described by Gaylor et al. [1998]. Further the benchmark dose lower bound (BMDL) can be computed using profile likelihoods, which are described by Crump and Howe [1985]. For simplicity in the calculation, we compute the male and female BMDL at the nominal level of $\alpha = 0.025$, which implies a combined nominal coverage $\alpha = 0.05$.

This model provides a method to examine the dose-response relationships based on all tumors using all available rat data for fine and ultrafine TiO_2 [Lee et al. 1985; Muhle et al. 1991; Heinrich et al. 1995]. Table A–3 provides the BMD (BMDL) excess risk estimates using this method on the pooled data for the response of either all-tumors or lung tumors excluding cystic keratinizing squamous lesions for the logistic regression model, which is one of the models in the BMDS model suite and which was selected based on the statistical properties of the logistic regression model that were amenable to developing this modified method (as discussed above). Table A–4 provides a comparison of the BMD (BMDL) excess risk estimates for all-tumors in male and female rat data, fit separately to each of the models in the BMDS model suite [EPA 2003].

Table A–1. Goodness of fit of logistic regression models to pooled rat data of lung tumor proportion and TiO$_2$ dose (as retained particle mass or surface area in the lungs) in rats after 24-month exposure[*]

Dose metric	Tumor response	Degrees of freedom	P-value (dose only model)	Degrees of freedom	P-value (dose and sex terms)
Surface area (m^2/g lung)	All tumors	10	0.056	8	0.29
Mass (mg/g lung)		10	< 0.0001	8	< 0.0001
Surface area (m^2/g lung)	No keratinizing cysts	10	0.50	8	0.62
Mass (mg/g lung)		10	< 0.0001	8	< 0.0001

[*]Pearson test for lack of fit. In the model with both dose and sex terms, the slopes and intercepts are averaged for the male/female combined average data from Muhle et al. [1991]. Rat data are from two studies of fine TiO$_2$ [Lee et al. 1985; Muhle et al. 1991] and one study of ultrafine TiO$_2$ [Heinrich et al. 1995] (12 data points total).

Table A–2. Tests for heterogeneity of rat sex or study/strain in dose-response relationship, based on likelihood ratio tests

Test[*]	Tumor response	Degrees of freedom	P-value	Heterogeneity
Rat sex (male vs. female)[†,‡]	All lung tumors	2	0.012	Yes
	No keratinizing cysts	2	0.14	No
Study/strain[†,§]	All lung tumors	4	0.46	No
	No keratinizing cysts	4	0.44	No
Ultrafine vs. fine (in females)[¶,**]	All lung tumors	2	0.66	No
	No keratinizing cysts	2	0.22	No

[*]Null model includes two terms: intercept and slope × surface area dose (m^2/g lung).
[†]Data include Lee et al. [1985] (male, female); Heinrich et al. [1995] (female); and Muhle et al. [1991] (male-female average)—12 data points total.
[‡]Full model includes four terms: separate intercepts and slopes for male and female rats (male-female average data were included assigned a value of 0.5 each for male and female indicators).
[§]Full model includes six terms: intercept and slope from null model (for comparison group), and separate intercept and slope terms for each of the other two study/strains.
[¶]Data include females from Lee et al. [1985] and Heinrich et al. [1995]—6 data points total.
[**]Full model includes four terms: intercept and slope from null model (for comparison group), and separate intercept and slope terms for the other group.

Table A–3. All tumors or lung tumors excluding cystic keratinizing squamous lesions: Logistic (sex-adjusted) model used to estimate benchmark dose (BMD) and lower 95% confidence limit (BMDL) estimates—expressed as TiO_2 particle surface area in the lungs (m^2/g)—in pooled rat data (males, female, and male-female average)[*]

Rat sex	DF	P-value (for lack of fit)	BMD (BMDL) by excess risk level	
			1/10[†]	1/1000[‡]
Tumors excluding cystic keratinizing squamous lesions				
Male	8	0.73	1.07 (0.81)	0.011
Female			1.04 (0.93)	0.010
All tumors				
Male	8	0.35	1.01 (0.78)	0.010
Female			0.85 (0.75)	0.0085

[*]Data are from two studies of fine TiO_2 [Lee et al. 1985; Muhle et al. 1991] and one study of ultrafine TiO_2 [Heinrich et al. 1995].
[†]Estimated directly from model.
[‡]Estimated from linear extrapolation of BMD and BMDL at 1/10 excess risk level.

Table A–4. All tumors: Benchmark dose (BMD) and lower 95% confidence limit (BMDL) estimates—expressed TiO$_2$ particle surface area in the lungs (m^2/g)—by model fit separately to male and female rat data

Model (BMDS 2003)	MALE rats [Lee et al. 1985]				FEMALE rats [Lee et al. 1985; Heinrich et al. 1995]			
	P-value (for lack of fit)	BMD (BMDL) by excess risk level			P-value (for lack of fit)	BMD (BMDL) by excess risk level		
		1/10*	1/1000*	1/1000†		1/10*	1/1000*	1/1000†
Gamma	0.51	1.11 (0.65)	0.54 (0.0062)	0.011 (0.0065)	0.20	0.76 (0.54)	0.20 (0.038)	0.0076 (0.0054)
Logistic	0.64	1.00 (0.82)	0.026 (0.018)	0.01 (0.0082)	0.15	0.86 (0.77)	0.050 (0.027)	0.0086 (0.0077)
Multistage	0.80	1.05 (0.65)	0.22 (0.0062)	0.010 (0.0065)	0.30	0.65 (0.51)	0.063 (0.0080)	0.0065 (0.0051)
Probit	0.62	0.98 (0.78)	0.023 (0.015)	0.0098 (0.0078)	0.24	0.79 (0.70)	0.044 (0.023)	0.0079 (0.0070)
Quantal-linear	0.40	0.87 (0.54)	0.0083 (0.0051)	0.0087 (0.0054)	0.068	0.37 (0.30)	0.0035 (0.0028)	0.0037 (0.0030)
Quantal-quadratic	0.73	0.98 (0.78)	0.096 (0.076)	0.0098 (0.0078)	0.30	0.65 (0.58)	0.063 (0.057)	0.0065 (0.0058)
Weibull	0.52	1.15 (0.65)	0.66 (0.0027)	0.012 (0.0065)	0.16	0.76 (0.52)	0.13 (0.024)	0.0076 (0.0052)

*Estimated directly from each model (in multistage, degree of polynomial: 3rd, male; 2nd, female).
†Estimated from linear extrapolation of BMD and BMDL at 1/10 excess risk level.

APPENDIX B: Threshold Model for Pulmonary Inflammation in Rats

A threshold model (i.e., piecewise linear or "hockeystick") was examined for its ability to adequately represent TiO_2-induced pulmonary inflammation in rat lungs [Tran et al. 1999; Cullen et al. 2002; Bermudez et al. 2002; Bermudez et al. 2004]. As described in Section 4.3.1.2, the TiO_2 pulmonary inflammation data from the Tran et al. [1999] and Cullen et al. [2002] studies could be fitted with a piecewise linear model which included a threshold parameter, and the threshold parameter estimate was significantly different from zero at a 95% confidence level. However, the fine and ultrafine TiO_2 pulmonary inflammation data from the Bermudez et al. [2002] and Bermudez et al. [2004] data sets provided no indication of a nonzero response threshold and were not consistent with a threshold model. The piecewise linear modeling methodology is detailed below.

In modeling pulmonary inflammation (as neutrophilic cell count in BAL fluid) in rat lungs, the response was assumed to be normally distributed with the mean response being a function of the dose and the variance proportional to a power of the mean. Thus for the ith rat given the dose d_i, the mean neutrophilic cell count would be $\mu_{pmn}(d_i)$ with variance $\alpha(\mu_{pmn}(d_i))^\rho$, where μ_{pmn} is any continuous function of dose, α is a proportionality constant, and ρ represents a constant power. The mean response was modeled using a variety of functions of dose; these functions were then used to estimate the critical dose at which the mean neutrophil levels went above the background. For the continuous functions that did not include a threshold parameter, this critical level was found using the BMD method [Crump 1984] and software [EPA 2003]. For purposes of calculation, the BMD was defined as the particle surface area dose in the lungs associated with $\mu_{pmn}(d_i)$ corresponding to the upper 5th percentile of the distribution of PMN counts in control rat lungs.

For the piecewise linear model, which is a threshold model, we assumed no dose-response, and thus no additional risk, above background prior to some critical threshold γ. For points beyond the threshold, the dose-response was modeled using a linear function of dose. For example,

$$\mu_{pmn}(d_i) = \begin{cases} \beta_0 & d_i < \gamma \\ \beta_0 + \beta_1(d_i - \gamma) & d_i \geq \gamma \end{cases}.$$

As the parameter γ is an unknown term, the above function is nonlinear and is fit using maximum likelihood (ML) estimation. Very approximate (1-α)% CIs can be found using profile likelihoods [Hudson 1966]. As the confidence limits are only rough approximations, the limits and significance of the threshold can be cross validated using parametric bootstrap methods [Efron and Tibshirani 1998].

APPENDIX C: Comparison of Rat- and Human-Based Excess Risk Estimates for Lung Cancer Following Chronic Inhalation of TiO$_2$

As described in Chapter 2, the epidemiologic studies of workers exposed to TiO$_2$ did not find a statistically significant relationship between the estimated exposure to total or respirable TiO$_2$ and lung cancer mortality [Fryzek et al. 2003; Boffetta et al. 2004], suggesting that the precision of these studies for estimating excess risks of concern for worker health (e.g., ≤1/1000) may be limited. The exposure data in Fryzek et al. [2003] were based on the total dust fraction, whereas respirable dust data were used to estimate exposures in Boffetta et al. [2004]. Neither study had exposure data for ultrafine particles. Chronic inhalation studies in rats exposed to fine [Lee et al. 1985] and ultrafine TiO$_2$ [Heinrich et al. 1995] showed statistically significant dose-response relationships for lung tumors (Chapter 3). However, the rat lung tumor response at high particle doses which overload the lung clearance has been questioned as to its relevance to humans [Watson and Valberg 1996; Warheit et al. 1997; Hext et al. 2005]. Recent studies have shown that rats inhaling TiO$_2$ are more sensitive than mice and hamsters to pulmonary effects including inflammation [Bermudez et al. 2002, 2004], although the hamsters had much faster clearance and lower retained lung burdens of TiO$_2$ compared to rats and mice. Because of the observed dose-response data for TiO$_2$ and lung cancer in rats, it is important to quantitatively compare the rat-based excess risk estimates with excess risk estimates derived from results of the epidemiologic studies.

The purpose of these analyses is to quantitatively compare the rat-based excess risks of lung cancer with results from the human studies. If the sensitivity of the rat response to inhaled particulates differs from that of humans, then the excess risks derived from the rat data would be expected to differ from the excess risks estimated from the human studies. The results of the comparison will be used to assess whether or not the observed differences of excess risks have adequate precision for reasonably excluding the rat model as a basis for predicting the excess risk of lung cancer in humans exposed to TiO$_2$.

Methods

Excess risk estimates for lung cancer in workers were derived from the epidemiologic studies (Appendix D) and from the chronic inhalation studies in rats [Heinrich et al. 1995; Lee et al. 1985; Muhle et al. 1991]. These excess risk estimates and associated standard errors were computed for a mean exposure concentration of 2.4 mg/m^3 over a 45-year working lifetime.

This exposure concentration was selected to correspond to a low value relative to the rat data (which is also the NIOSH REL, see Chapter 5).

Excess risks were derived from the rat data based on the three-model average procedure described in Chapter 4. The Model Averaging for Dichotomous Response (MADR) software used for the model average risk estimation was modified to output the model coefficients and model weights associated with each model fit in a 2,000-sample bootstrap, based on the rat data. The model coefficients and weights were then used to construct the distribution of risk estimates from rats exposed to the equivalent of a lifetime occupational exposure to 2.4 mg/m³ fine TiO_2. The rat-equivalent exposures were estimated by using the MPPD2 model to estimate the human lung burden associated with a 45-year occupational exposure to 2.4 mg/m³ fine TiO_2. This lung burden (2545 mg TiO_2) was then extrapolated to rats following the steps described in Chapter 4 for extrapolating from rats to humans, but in reverse order. This procedure yielded an estimated rat lung burden of 6.64 mg TiO_2. Since the rat-based, dose-response modeling was based on a particle surface area dose metric, the rat lung burden was then converted to 0.0444 m² particle surface area per gram of lung tissue, assuming a specific surface area of 6.68 m²/g TiO_2. The 2,000 sets of model equations from the model average were then solved for a concentration of 0.0444 m²/g lung tissue, yielding a bootstrap estimate of the distribution of the excess risk associated with exposure to TiO_2.

Excess risks were estimated from each of the two worker cohort studies, using two different methods for each. For the cohort studied by Boffetta et al. [2004], two different values for representing the highest cumulative exposure group were separately assumed; and for the cohort studied by Fryzek et al. [2003], two different exposure lags (no lag, 15-year lag) were separately used.

Results

Table C shows the rat-based maximum likelihood estimate (MLE) and 95% upper confidence limit (UCL) of excess risks for lung cancer and the human-based 95% UCL on excess risk from exposure to TiO_2. There is consistency in the estimates of the 95% UCL from these two independent epidemiologic studies at the exposure concentration evaluated for both studies, 2.4 mg/m³ (Boffetta: 0.038 and 0.053; Fryzek: 0.046 and 0.056). The rat-based MLE and 95% UCL excess risk estimates for fine TiO_2 exposures are lower than the 95% UCL risk estimates based on the human studies in Table C. This result suggests that the rat-based risk estimates for fine TiO_2 are not inconsistent with the human risk estimates derived from the epidemiological studies.

Discussion

These two epidemiologic studies are subject to substantially larger variability than are the rat studies. The results of the epidemiologic studies of TiO_2 workers by Fryzek et al. [2003] and Boffetta et al. [2003, 2004] are consistent with a range of excess risks at given exposures, including the null exposure-response relationship (i.e., no association between the risk of lung cancer and TiO_2 exposure) and an exposure-response relationship consistent with the low-dose extrapolations from the rat studies (based on the method used, a three-model average as described in Chapter 4). Both the MLE and 95% UCL excess risk estimates from the rat studies were lower than the 95% UCL from the human studies for fine TiO_2.

In conclusion, the comparison of the rat- and human-based excess risk estimates for lung cancer suggests that the rat-based estimates for exposure to fine TiO_2 particles are not inconsistent with those from the human studies. Therefore, it is not possible to exclude the rat model as an acceptable model for predicting lung cancer risks from TiO_2 exposure in workers.

Table C. Comparison of rat-based excess risk estimates for lung cancer from TiO_2 (using a model average procedure) with the 95% upper confidence limit (95% UCL) of excess risk of lung cancer in workers, at 2.4 mg/m³ exposure concentration, for a 45-year working lifetime[*]

TiO_2 mean concentration (mg/m³) over 45-year working lifetime	Human-based excess risk (95% UCL): two different estimates from Boffetta et al. [2003, 2004]	Human-based excess risk (95% UCL): two different estimates from Fryzek et al. [2003]	Rat-based excess risk estimate for fine TiO_2 MLE	95% UCL
2.4	0.038[†]	0.056[§]	0.000005	0.0031
	0.053[‡]	0.046[¶]		

[*]*Methods notes*: The value of 2.4 mg/m³ is a low value relative to the rat study, and is also the NIOSH REL. The MPPD human lung model [CIIT and RIVM 2002] was first used to estimate the lung burden after 45-years of exposure to a given mean concentration. The estimated retained particle mass lung burden was extrapolated from human to an equivalent particle surface area lung burden in rats, based on species differences in the surface area of lungs, and using the specific surface area value of fine TiO_2 (6.68 m²/g). The rat dose-response model (three-model average, see Chapter 4) was then used to estimate the excess risk of lung cancer at a given dose.
[†]From Boffetta et al. [2003, 2004]] assumed 78.1 mg-yr/m³ in highest cumulative exposure group (respirable TiO_2).
[‡]From Boffetta et al. [2003, 2004], assumed 56.5 mg-yr/m³ in highest cumulative exposure group (respirable TiO_2).
[§]From Fryzek et al. [2003, 2004a,b]; Fryzek [2004] unlagged model (total TiO_2).
[¶]From Fryzek et al. [2003, 2004a,b]; Fryzek [2004] model with 15-year lag (total TiO_2).

APPENDIX D: Calculation of Upper Bound on Excess Risk of Lung Cancer in an Epidemiologic Study of Workers Exposed to TiO_2

Results from two epidemiologic studies [Fryzek 2004; Fryzek et al. 2003, 2004a; Boffetta et al. 2003, 2004] were used to compute the upper bound estimates of excess lung cancer risk. The excess risks for lung cancer corresponding to the upper limit of a two-sided 95% CI on the RR associated with cumulative exposure to total TiO_2 dust in U.S. workers were based on results supplied by Fryzek [2004] for Cox regressions fitted to cumulative exposures viewed as a time-dependent variable. The provided results include the coefficients and standard errors for the continuous model for cumulative exposure [Fryzek 2004]. For a study of United Kingdom and European Union workers exposed to respirable TiO_2 [Boffetta et al. 2004], excess risks for lung cancer were not available and therefore were derived from the results provided in a detailed earlier report Boffetta et al. [2003], as follows. The excess risk estimates computed from each of these epidemiologic studies were then used in Appendix C for comparison to the rat-based excess risk estimates for humans (Chapter 4).

Methods

Categorical results on exposure-response are reported in Tables 4–1 (SMRs) and Table 4–2 (Cox regressions) of Boffetta et al. [2003].

There are four categories—i.e., 0–0.73, 0.74–3.44, 3.45–13.19, 13.20+ ($mg/m^3 \cdot yr$)—in these results, and the maximum observed exposure is 143 $mg/m^3 \cdot yr$ (Table 2–8 of Boffetta et al. [2003]). Hence, the midpoints of the categories are 0.365, 2.09, 8.32, 78.1 $mg/m^3 \cdot yr$. The value of the highest category depends on the maximum observed value and is subject to considerable variability. An alternate value for this category is 56.5 $mg/m^3 \cdot yr$. This value is based on estimating the conditional mean cumulative exposure given that the exposure exceeds 13.20 using the lognormal distribution that has median 1.98 and 75th percentile equal to 6.88 based on results in Table 2–8 (*Overall*) of Boffetta et al. [2003]. Results are generated using both 78.1 and 56.5 $mg/m^3 \cdot yr$ to represent the highest exposure group. The SMRs reported in Table 4–1 were modeled as follows:

E[SMR] = Alpha*(1+Beta*CumX) where SMR = Y/E is the ratio of the observed to the expected count.

=> E[Y] = Alpha*(1+Beta*CumX)*E fitted to observed counts (Y) by iteratively reweighted least squares (IRLS) with weights proportional to 1/E[Y].

Notes: Beta describes the effect of cumulative exposure, CumX, and Alpha allows the cohort to differ from the referent population under unexposed conditions.

The estimators of Alpha and Beta are based on iteratively reweighted least squares with weights proportional to the reciprocal of the mean. Although these estimates are equivalent to Poisson regression MLEs, the observed counts are not strictly Poisson. This is due to the adjustments made by Boffetta et al. [2003] for missing cause of death arising from the limited time that German death certificates were maintained. The reported observed counts are 53+0.9, 53+2.3, 52+2.7, 53+2.4 where 0.9, 2.3, 2.7, and 2.4 have been added by Boffetta et al. [2003] for missing cause of death that are estimated to have been lung cancer deaths. Invoking a Poisson regression model should work well given such small adjustments having been added to Poisson counts of 53, 53, 52, and 53. Hence, Alpha and Beta are estimated accordingly but their standard errors and CIs do not rely on the Poisson assumption; instead, standard errors were estimated from the data and CIs were based on the t distribution with 2 degrees of freedom.

A similar approach using the results of Table 4–2 was not attempted since these categorical RR estimates are correlated and information on the correlations was not reported by Boffetta et al. [2003].

Results

Results based on modeling the SMRs in Table 4–1 of Boffetta et al. [2003] with a linear effect of cumulative exposure are presented in Table D–1. These results are sensitive to the value used to represent the highest cumulative exposure category, particularly the estimate of the effect of exposure. However, zero is contained in both of the 95% CIs for Beta indicating that the slope of the exposure-response is not significant for these data.

Estimates of excess risk based on application of the results given in Table D–1 to U.S. population rates using the method given by BEIR IV [1988] appear in Table D–2.

Discussion

The exposure assessment conducted by Boffetta et al. [2003] relies heavily on tours of the factories by two occupational hygienists who first reconstructed historical exposures without using any measurements (as described in Boffetta et al. [2003]; Cherrie et al. [1996]; Cherrie [1999]; Cherrie and Schneider [1999]). The sole use of exposure measurements by Boffetta et al. [2003] was to calculate a single adjustment factor to apply to the previously constructed exposure estimates so that the average of the measurements coincided with the corresponding reconstructed estimates. However, Boffetta et al. [2003] offer no analyses of their data to support this approach. Also, the best value to use to represent the highest exposure interval (i.e., *13.20* + mg/m³•yr) is not known and the results for the two values examined suggest that there is some sensitivity to this value. Hence, these upper limits that reflect only statistical variability are likely to be increased if the effects of other sources of uncertainty could be quantified.

Table D–1. Results on Beta from modeling the SMRs reported in Table 4–1 of Boffetta et al. [2003] for the model, E[SMR] = Alpha*(1+Beta*CumX)

Value representing highest CumX	Beta* estimate	Approx std error	Approximate 95% confidence	Limits
78.1	0.000044	0.00163	-0.00697	0.00706
56.5	0.000109	0.00229	-0.00975	0.00996

*Beta is the coefficient for the effect of 1 mg/m^3•yr cumulative exposure to respirable TiO$_2$ dust.

Table D–2. Lifetime excess risk after 45 years of exposure estimated by applying the above UCLs on Beta and the linear relative rate model of lung cancer to U.S. population rates*

Occupational exposure (8-hr TWA respirable mg/m^3)	Background risk (Ro)	Beta=0.000044 excess risk[†] (Rx-Ro)	UCL=0.00706 excess risk[†] (Rx-Ro)	Beta=0.000109 excess risk[‡] (Rx-Ro)	UCL=0.00996 excess risk[‡] (Rx-Ro)
0.0	0.056	0	0	0	0
1.5		0.0002	0.024	0.0004	0.033
5.0		0.0005	0.076	0.0012	0.11
15.0		0.0015	0.21	0.0037	0.27

*Based on the method given by BEIR IV using U.S. population rates given in Vital Statistics of the U.S. 1992 Vol II Part A [NCHS 1996]. Occupational exposure from age 20 through age 64 and excess risks subject to early removal by competing risks are accumulated up to age 85.

[†]Value representing the highest exposure category is 78.1 mg/m^3 yr based on the midpoint of the interval [13.20, 143].

[‡]Value representing the highest exposure category is 56.5 mg/m^3 yr based on the conditional mean given exposures greater than 13.20 using the conditional distribution derived from the lognormal distribution having median and 75th percentiles equal to 1.98 and 6.88 mg/m^3 yr, respectively.